LOW~ RESISTANCE BOATS

LOW~ RESISTANCE BOATS

Build 24 Boats that Move Easily Through the Water

Thomas Firth Jones

International Marine Publishing
Camden, Maine

Some parts of this book have appeared, in different form, in *Multihulls* and *Small Boat Journal.*

Published by International Marine

10 9 8 7 6 5 4 3 2 1

Library of Congress Cataloging-in-Publication Data

Jones, Thomas Firth, 1934–
    Low resistance boats / Thomas Firth Jones.
       p. cm.
    Includes index.
    ISBN 0–87742–284–2
    1. Boatbuilding. I. Title.
    VM321.J66 1991   91–24941
    623.8′2––dc20   CIP

Questions regarding the content of this book should be addressed to:

International Marine Publishing
P.O. Box 220
Camden, ME 04843

Typeset by A & B Typesetters, Inc., Bow, NH
Printed by Arcata Graphics, Fairfield, PA
Design by Joyce C. Weston
Illustrated by the author
Edited by J.R. Babb and Keith Lawrence
Production by Janet Robbins

for Carol

Contents

Introduction

This book describes 24 boats I've built or designed. None is more than 29 feet long, and many are small and light enough to be carried on one shoulder. Thirteen are sailboats, seven are powerboats, and four are paddle or rowing boats. Half have plywood hulls, and the rest have other kinds of wooden or fiberglass/foam sandwich hulls. Their uses also vary widely: The *El Toro* is a tiny one-design racing sailboat; *Vireo* and the *Hummingbird* are ocean-crossing multihulls; and *Puxe* is a semi-displacement launch intended for sheltered water.

The discussion is not highly technical, but does assume that the reader already knows something about boats. He (women are included, whenever *he* is used in this impersonal sense) may be looking for a design to build, different solutions to old building or designing problems, or just some boats to imagine over a few wintry evenings.

Two of the boats—the arc-bottomed kayak, and the planing garvey—use methods unorthodox enough to warrant a full and minute description of the building process. However, when it comes to conventional plywood construction, cold-molding, carvel and clinker planking, or foam sandwich, a number of good primers are already in print. In discussing these techniques, I have limited myself to describing what can be done that differs from standard practice, and to warning of certain pitfalls. I have tried to give a feel for each technique—how long it takes; how pleasant or unpleasant it is to do; what strength, weight, and quality of finish are likely to result. Thus the reader who has, for example, only built plywood boats and is thinking of clinker or foam sandwich for his next project, can get an idea whether or not he wants to tackle it.

Scampi is the only boat in this book designed to be trailered. However, all but the multihulls in the last chapter could be trailered. The paddling boats and several of the daysailers can be readily car-topped. Some of the other boats would require special trailers, but good trailer-builders are ingenious. Moreover, it is worth remembering that boats are water vehicles, and a boat that fits a stock trailer may give less pleasure in her proper element.

My wife, Carol, and I use some of these boats, some were commissioned by customers, and some were built on speculation. Most of them—and nearly all of them, where I had any control over the design—were intended to move through the water as easily as possible, consistent with their function and cost. Of all the forms of locomotion, boats are the slowest, and there are only two ways to achieve speeds that make them interesting to use: one is high power, and the other is low resistance. High power is expensive, and will become increasingly so, out of proportion to inflation. It also strains the structure, which must therefore be stronger, and thus heavier, requiring more power and more expense—an unending cycle. Low resistance is the product of good design, having more to do with mathematics than with lines. Considering that this math is not nearly as complicated as algebra, it is astonishing how much of it was not discovered until this century. More than any other factor, low resistance is the product of light weight.

Function and cost make boat design complex, not the simple math of low resistance. Much of this can be solved if the buyer or builder will think honestly about his requirements and carefully define his intended use of the boat. Noah built an ark, but after the deluge, he and his wife found her quite a handful, so they beached her on Mount Ararat and spent their golden years playing bingo. Our times are more affluent than Noah's, and too many of us imagine that even if the elephant and the kangaroo will only be aboard once, they'll need private staterooms and heads. A smaller boat of good materials will give better service and satisfaction over time than a larger, shoddier one. Most boats are needlessly large.

Low resistance is equally desirable in power, sail, and paddling boats. Beauty can perhaps be integrated with it, but is no substitute for low resistance, and this is equally true whether the beauty is functional or quaint. For myself, I prefer quaint looks in a monohull and functional looks in a multi. These judgments are subjective, however, and I wouldn't attempt to defend them. For power, I'd usually rather sail, and Carol would usually rather paddle. However, because of its great range and predictable schedule, our motor launch is the boat we take out most often.

On our first visit to the Azores, we were sitting in the saloon of the last *Gypsy Moth*, admiring her many intricacies. Among those were the twin knotmeters, each one to be used on a different tack, and the cut-glass vase in Lady Chichester's cabin, which was always kept full of flowers even when she wasn't aboard. To Giles Chichester, who had inherited the boat, I said that it might be more satisfying to use a boat of your own design and construction, even if it wasn't as good as someone else's. He took his pipe from his mouth, as if to speak, but only answered by shaking his head. Though I was crushed at the time, I soon recovered, and I'm still pretty sure he was wrong.

A hundred years ago, Americans often made their own boats, as well as their own houses, clothing, and furniture. Today, many Americans don't even cook their own food, and don't know how. The free time that this gives them is partly spent in a second job, making enough money to pay the people who cook and build boats for them, and partly spent in selecting and buying merchandise. This is a hollow life, with shallow and fleeting satisfactions, ever-increasing dissatisfaction, and no possibility of personal fulfillment. The greatest benefit of work is not the wage or even the product, but the chance to learn and grow.

I was nearly 40 years old when I built my first boat, and nearly 50 when I opened my own shop. True, I had considerable experience using boats built by other people, and had made my living at carpentry most of my working life. Nevertheless, there was much to learn, and there still is. Join me.

My work is building boats, designing them, and selling plans. Serious inquiries about any of these things are welcome. However, before you write, please ask yourself how feasible your project is, how seriously you are interested, and how much time it may take me to answer you. When wind and tide are right, I really do like to be out on the water.

Tom Jones
Jones Boats
Box 391
Tuckahoe, NJ 08250

1. Flat-Water Kayaks

The kayak was invented by the Eskimo, who used it in the open sea. Because of this he needed a decked boat, and often he needed to right it after capsize—the famous "Greenland roll." Today, at kayak slaloms, one often sees the waiting contestants roll their boats three or four times, just for the fun of it. It must have been less fun for the Eskimo in the frigid Arctic Sea.

For slaloms or for sea kayaking, a decked boat is still best. For flat water, some decking can give lower freeboard and windage, which can be a powerful source of resistance in low-powered craft. But decks weigh something, and they restrict the skipper's mobility. When it's time to take off the sweater, where to put the paddle? And how to reach that Thermos, jiggling so invitingly below deck? Where to stow that fine-looking milk crate stuck in the weeds over there? For flat water, a kayak had better be an open boat.

Dread Bob and the Kingston Wailer

These boats were built soon after reggae singer Bob Marley "went into himself," as the Rastas say. They were named in his honor, and he would have enjoyed their simplicity. They are really two different sizes of the same boat, and either can be built in 50 hours. They give the builder a taste of cold molding, and an introduction to designing a hull to suit the weight it must carry. Let's begin by building the *Wailer*.

Topsides are 8-inch rips of 1/4-inch plywood, though 3/16-inch ply could just as easily be used, and would save about 3 pounds in the completed boat. Topsides are straight-edged; rocker and sheer are provided by the flare. The panels can be scarfed to 12-foot lengths, or they can be butt-blocked after chine logs and sheer clamps are attached. If butt-blocked, the blocks on port and starboard sides should be the same distance from the ends, or the boat will be asymmetrical.

A pattern is made for the bow and stem curve, to give about 2 inches overhang. This can be simply the arc of a circle, though a diminishing-radius curve (with more curve at bottom than top) is prettier. In cutting, be sure that all four ends are exactly alike, and topsides panels are exactly the same length.

Chine logs and sheer clamps must be continuous 12-foot lengths of

1

Figure 1–1. The Kingston Wailer.

lumber, not butt-jointed. They are glued to the topsides, and may be clamped, tacked, or stapled while the glue sets. Staples may be steel, and driven over coarse string to make extraction easier. After the fastenings are removed, chine logs and sheer clamps are cut roughly to the half-angle of the bow (16 degrees) and topsides are sprung around the temporary midships mold. The boat is now on sawhorses, and a hand saw is run down the centerline at bow and stern—two or three times until the fit is satisfactory. Do not try to bring these ends to razor edges, because they will need one light layer of glass tape after the bottom is on. Fiberglass needs a reasonable radius to conform, and will not lie down around a sharp edge.

Run a taught string from bow to stern to be sure that the boat is symmetrical. Screw the chine logs and sheer clamps together. Cut the inner stem and stern posts from 3/4-inch scrap, bevel them, and glue and clamp them in place. Alternatively, glass tape or epoxy putty may be used to join the insides of the plywood at bow and stern. The keelson is fitted (it will need to be slightly longer than a straight-line distance) and attached with one long screw through the topsides at each end. No need to bother with glue.

A camber board of 42-inch radius is now laid over the chine logs and keelson and moved along to show the changing bevel to be cut with a plane. Bottom planking is now begun, using 1/8-inch cedar 2 to 4 inches wide, which may be purchased or ripped on a table saw. All knots should be cut to waste, as should all sapwood, which has no place in any part of any boat's structure because it rots so quickly, no matter what the wood species. Planking is laid at 45 degrees to the keelson, beginning amid-

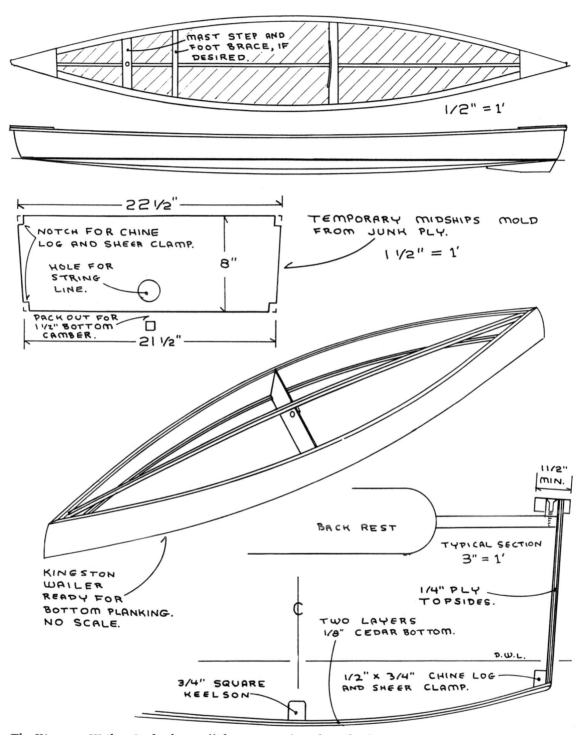

MAST STEP AND FOOT BRACE, IF DESIRED.

1/2" = 1'

22 1/2"

NOTCH FOR CHINE LOG AND SHEER CLAMP.

HOLE FOR STRING LINE.

8"

TEMPORARY MIDSHIPS MOLD FROM JUNK PLY.

1 1/2" = 1'

PACK OUT FOR 1 1/2" BOTTOM CAMBER.

21 1/2"

1 1/2" MIN.

BACK REST

TYPICAL SECTION
3" = 1'

KINGSTON WAILER READY FOR BOTTOM PLANKING. NO SCALE.

1/4" PLY TOPSIDES.

TWO LAYERS 1/8" CEDAR BOTTOM.

D.W.L.

3/4" SQUARE KEELSON

1/2" X 3/4" CHINE LOG AND SHEER CLAMP.

The Kingston Wailer. *Scale these off the page at ½-inch to the foot.*

3

ships and working forward and aft. Each plank is glued and tacked or stapled to the framing, and every fourth or fifth plank must be scribed to its neighbor so it will lie down properly. Planks are best run past the chine logs, and cut off in batches later. This work is time-consuming, and anyone considering a bigger cold-molded project (a 50-foot schooner, for example, made with 12 layers of 1/8-inch veneer) should first try a project this size to test his own patience. I figure one-half hour per square foot per layer, not counting the time it takes to make the veneer.

In the second layer of veneer, fastenings must be extracted, although in the first, staples or tacks can be left in place if they are non-ferrous. They will not bleed through or work out through the second layer. Bell Telephone uses endless Monel staples and many other good boatbuilding materials, but, of course, I don't know how to get hold of them. When the first layer is completed, the edges where the planks butt each other should be painted with glue thin enough to penetrate the seam and join them together. Then the whole bottom is sanded, bearing in mind that cedar cuts quicker than glue. The second layer of planking is begun, again amidships, at right angles to the first.

Liberal quantities of glue should be used to join the second layer to the first, and in some places it may be necessary to run a screw through to pull the layers together. Silicon spray is a good parting agent to put on such a screw. Waxed paper is also an excellent parting agent for all glues, and for polyester resin as well. If a plank really doesn't want to lie down, it may need scribing, or just discarding in favor of a more tractable piece of wood. You'll get a feel for it as you go along.

Another way to make the second layer of veneers lie down on the first is vacuum bagging. This is a complicated procedure, and needs much equipment and set-up time. Shops that vacuum bag regularly and have developed systems for setting up the equipment and keeping it clean, say it's the best. It is certainly better for fiberglass work with polyester resin than for gluing work, because polyester goes off so quick. The lurking danger with vacuum bagging is vacuum failure, from bags leaking or pumps quitting. The resultant mess is so awful that often it's better just to go away and set up a new shop in a new location. At any rate, it isn't worthwhile to set up for vacuum bagging for a project the size of the *Wailer.*

When the bottom is finished, to keep the boat from losing some of its shape, the backrest spreader should be glued and screwed in before the mold is removed. It goes a foot aft of midships. If a sailing rig is desired, best stick the partners in now. A foot brace, which does contribute to speed and ease of paddling, is best left out until the boat is tested in the water, and best not glued for some time, if ever, because its distance from the backrest is critical and individual.

The sheer is packed out to at least 1½-inch width, and the skeg, ¾ inch × 3 inches × 18 inches, is shaped for water flow, glued, and screwed through the keelson. The breasthook decks are then glued and fastened down. They should be about a foot long, of ¼-inch ply. The backrest is also ¼-inch ply, and may be curved for looks. For comfort, it should slope aft 15 degrees. Also for comfort, a lifesaving cushion will be needed to pad the keelson. The *Kingston Wailer* is done. Paint her and try her out.

Designing a Kayak

The *Wailer* was designed for a 100-pound paddler, and the calculations were fairly simple:

clothed paddler	105 lbs.
boat	25 (this proved correct)
paddle, cushion, etc.	10
total	140 lbs.

As water weighs 64 pounds per cubic foot, the boat needed about 2.2 cubic feet of hull below the waterline to float the load. This is true displacement. Too often in advertisements boat weight is called displacement, but a boat that displaced 25 pounds would suit only the owl and the pussycat.

It is safe to assume, in a boat of this type with pointed ends and no bumps, hollows, or flats, that the prismatic coefficient will be about .55. That is to say, if a prism has the same waterline length as the *Wailer* (11½ feet), and the same section throughout that the *Wailer* has amidships, and the prism displaces 4 cubic feet, then the *Wailer* will displace 2.2 cubic feet, or 55 percent of what the prism does. Prismatic coefficient is a ratio describing the fineness of the ends.

Therefore the midships section must be 4 cubic feet divided by 11½ feet, which is .348 square feet, or 50 square inches. The arc bottom was chosen to reduce wetted surface, and to make the work interesting enough to be worth doing. The *Wailer* could have a flat bottom, but she would need more rocker to carry the same load, and she'd have 8 percent more wetted surface. This is the primary source of resistance at the speed a cruising kayak goes. She could be built in a very few hours, but what to do after that? Some people build another, and soon have a whole flotilla of not-very-good boats. But certainly if you need a boat for an urgent purpose, don't build it cold-molded.

The 22-inch bottom width is narrow enough for a cruising boat, though

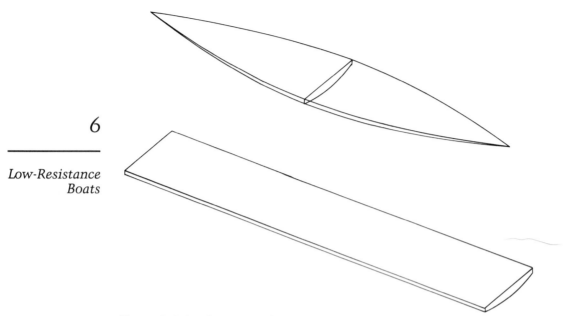

Figure 1–2. Underwater volumes: Kingston Wailer *(top) and prism (bottom).*

racers should be narrower. It gives a length-to-beam ratio of 6.25, which helps reduce wave-making—the secondary resistance factor. Eighth-inch veneers bend easily to an arc of 1½ inch in 22 inches, giving an area within the arc of 25 square inches (1½ × 22 × ¾). That leaves another 25 square inches to be found in the 22-inch-wide section above the arc, requiring a rocker of 1⅛ inches.

A little change in width at the top of the mold greatly changes the rocker of the boat. Using a string line through the hole in the mold, I arrived at the mold dimensions shown. They could easily be varied to suit a paddler of different weight, but there are limits to how much they can be changed.

The *Kingston Wailer* is not a good boat for a 170-pound paddler. One friend who tried it managed to get up the river and back, but said he never dared turn around to look behind him, for fear of capsizing. In addition, his extra weight sank the shaped part of the hull too deeply into the water, so that the intended shape was lost. He needed a bigger boat—wider for more stability, and longer to retain the sleek proportions. Wetted surface would be greater, but presumably the muscle-power to push it would be greater too. And because the boat would be longer, it would have a higher top speed.

Dread Bob was designed for a 160-pound paddler:

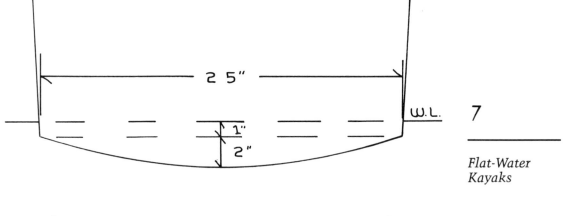

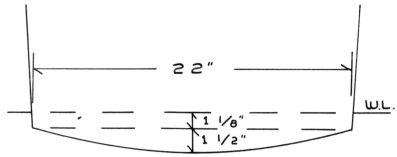

Figure 1–3. Midships sections: Dread Bob *(top) and the* Kingston Wailer *(bottom).*

clothed paddler	165 lbs.
boat	30 (again correct)
paddle, etc.	10
total	205 lbs.

Then, by the same kind of calculation as before:

$$\frac{205 \text{ lbs.} \times 144}{64 \text{ lbs.} \times .55 \text{ p.c.} \times \text{ft. W.L.}} = 62 \text{ square inches}$$

Cedar veneers bend easily to a 2-inch arc over a 25-inch bottom width, so with 37 square inches in the arc, only another 25 square inches are needed above it, or a rocker of one inch. The camber board was cut to a 40-inch radius. *Bob* was made with 9-inch topsides to give a little more security. Because the keelson was longer but no thicker and no more arched, it had to be temporarily supported at two other places to keep it from flattening out as the bottom veneers went on.

Variations in this calculation will produce an arc-bottomed kayak to suit any size paddler. A double kayak could be 16 feet by 30 inches, but much of the fun of kayaking is the absence of a command structure, so I'd rather have two singles. For a single not meant for racing, *Bob*'s dimensions are plenty big enough. Increasing either length or beam will increase wetted surface, and although it can be muscled into submission, a 14-footer will give more miles for the horsepower you are willing to put into it over a period of hours. To modify either *Bob* or the *Wailer* to suit a paddler of different weight, the best method is to increase or decrease the rocker with a slight change of the mold width at the sheer.

Kayak Power

Running upstairs against a stop watch, you can soon see the horsepower your legs generate. Running starts, which use the momentum of previous energy, are not allowed. One horsepower is the energy needed to lift 550 pounds one foot in one second, or 122 pounds 9 feet in 2 seconds, or 183 pounds the same distance in 3 seconds, etc. If you are young and strong, your legs may generate one and a trifle horsepower. But at a sustainable pace, they might generate one-tenth of a horsepower, and your arms might be half as strong as your legs. So kayaks are very low-powered craft, and they need to be very light and have good proportions for low resistance or they aren't any fun. A neighbor of ours built the famous L. Francis Herreshoff "double paddle canoe." Sixteen feet long, though only 13 feet 4 inches on the water, it has a cross-planked bottom, clinker topsides, and considerable deck. It looks just beautiful, but weighs 95 pounds, and he seldom uses it.

Figure 1–4. Wailer *under way. Carol's cross-legged posture is not the best for racing.*

Here on the Tuckahoe River, a local gun club sponsors an annual canoe and kayak race. There are five classes: racing canoes, two-man canoes, two-woman canoes, mixed canoes, and kayaks. The first two classes race five statute miles—up river, around an island, and home again. The other classes race a three-mile course, rounding a nearer island. The mass start is pandemonium, and well worth the trip to see. Soon after, it becomes clear that the best boats in the first two classes are moving away from the fleet. They are 18 or 20 feet on the water, and the crews use short, bent paddles and the brutal modern stroke that takes advantage of back as well as arm muscles. But the kayaks are not far behind, and at the finish even a mediocre kayaker has time to start on a beer before the first canoe arrives. I think kayaks are better boats than canoes, for a least three reasons: the motion is symmetrical, balanced, and less tiring; less energy is wasted moving the paddle when it isn't in the water; and the crew sits lower, allowing a narrower, sleeker shape. Canoes have the advantage of turning quicker, with bow and stern thrusters, in effect, so are easier managed in narrow streams and white water. I also prefer a canoe's looks, but as beauty is well known to be in the eye of the beholder, the less said about it, the better.

In 1990, the winning kayak covered three miles of the river in 31 minutes. He said he'd rather not do it again. According to formulas in Skene's *Elements of Yacht Design,* and presuming wind and tide canceled themselves out on this course, the 5.05 knot speed that he averaged with his 15-foot waterline required one-twentieth horsepower. Given this low power, paddle design becomes important too.

If kayak paddles had parallel blades—and the cheapest ones often do— to strike the water squarely, paddlers would still have to rotate them 45 degrees with each stroke. Usual practice is to arrange the blades at 90 degrees, so that the blade in the air has less wind resistance. The shaft is then grasped firmly in one hand and rotated in the other. This leads to the anomaly of right-handed and left-handed paddles, depending on

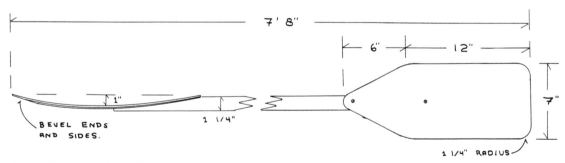

Figure 1–5. Kayak paddle.

which hand does the grasping. Some paddlers can pick up either kind and become accustomed in a few strokes. Others are so accustomed to one kind that when handed the other in the morning, they are still whiffing the water in the late afternoon.

Surprisingly, the paddle in the drawing has proven to be about the right length and blade area for paddlers of many different sizes. The shaft is 1¼-inch straight-grained fir closet rod, and the blades are laminated from two layers of ⅛-inch plywood or veneer over a form. If veneers are used, the grains should be somewhat diagonal and opposed. Cedar is strong enough if you're sure it will never touch bank or bottom, but fir or mahogany will take far more abuse. Back edges should be well beveled all around.

A paddle blade with squarer shoulders would drip less water into the boat, but we've found them cumbersome and inefficient. Drip cups are not necessary: in cruising, a short and welcome hesitation in the stroke when the blade is horizontal keeps most drips out. In racing, most drips come from the forehead, and damn the paddle. The curved blade, like the arc bottom of *Bob* and *Wailer*, is what makes paddle building worth the effort. Dick Young, who races our clinker boat on the Tuckahoe every year, tried it with a curved-blade paddle in 1990, and said it made a great difference. A genuinely cupped blade, curved in two directions, would have to be better yet; but it would be hard to make as light as a single curve, and a couple of ounces might be worth more than a perfect shape.

Sailing Rig

No matter how well a kayak paddles, I don't take kindly to paddling downwind. In very strong winds, a kayak will do about 2 knots on hull and body windage. Steering is achieved by holding the paddle at about shoulder height and rotating it 90 degrees, to let the wind push on one blade or the other. In lighter air a sail is desirable. A good kayak sail can be made from 2½ yards of light nylon, 60 inches wide, bought from the fabric store, not from the sailmaker. Cloth should not be porous; you shouldn't be able to breathe through it.

Make one diagonal cut, so that the luff is 6 feet and the leech is 7 feet. The cut-off makes the quarter-circle tablings at the corners. Their edges can be sealed with a hot knife, but the four sides of the sail should be hemmed. All lines are ⅛ inch and are simply sewn to the cloth without grommets. For the robands, sew both ends and slip them over the mast, which is 1¼-inch fir, 7½-feet long. The top 3 feet should taper. The sprit is ¾-inch square, 8½-feet long, tapered at both ends. Lines go through holes in the spars. The sail will be about 32 square feet.

When paddling, this rig is rolled up and tucked in beside the skipper.

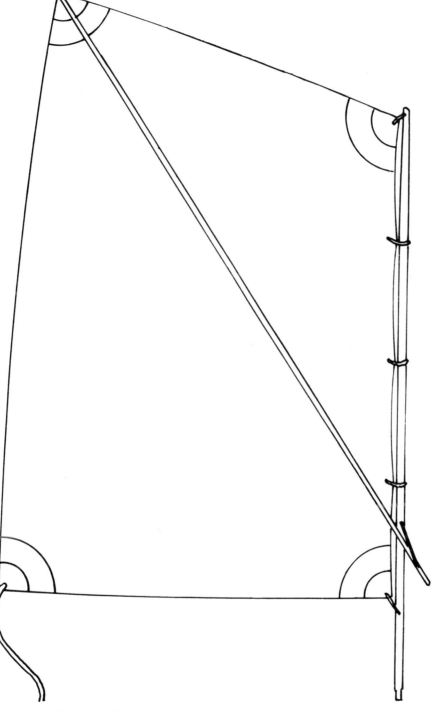

Figure 1–6. Kayak sailing rig.

When the wind comes fair, he unrolls it, slides forward, and steps it. Then he'd better be alert. The 8-foot sheet must be hand held at all times, not tied or cleated, and often it must be cast off. Neither *Bob* nor the *Wailer* has ever capsized under sail, but when winds are strong or water is cold, we leave the sailing rig in the shed. Carol doesn't like sailing a kayak in any weather, and has made me remove the step and partners from her boat. The rig is really too big for its narrow beam, and 20 square feet would be more appropriate.

But I do love sailing *Dread Bob*. It isn't fast or weatherly, and it will never tack (kayaks are too long to turn quickly, and they lack momentum), but it's endlessly interesting. The paddle is sometimes needed, for quick turns, but the real fun is to steer without it. Sheeting in and heeling to windward heads her off; easing the sheet and heeling to leeward heads her up. I have sometimes sailed *Bob* several miles on the winding, narrow Tuckahoe without touching paddle to water.

The mast is stepped well forward. Balance in a sailboat is desirable only for close-hauled courses, and harms steering with the wind aft. Ancient Egyptian boats, which drifted down the Nile and sailed back up, stepped the mast in the bow. Spinnakers, too, bring the center of effort far forward. Nevertheless, I did eventually trick *Bob* out with a daggerboard trunk.

Figure 1–7. Dread Bob *under sail.*

To avoid cutting the keelson, the trunk was slightly off center. The board was 8-inches wide and extended 12 inches below the hull. If lateral resistance is provided by an efficient shape, then one square foot of it is enough for each 40 to 50 square feet of sail area. The center of lateral resistance was under the sail's center of effort, as I did expect to close reach with the new set-up.

Bob went astonishingly faster with the daggerboard, even on a broad reach. Apparently, much of the sail's drive had been wasted in pushing the boat to leeward. Now it really would close reach, though the stretch of the nylon sail precluded windward work, and it still wouldn't tack. Sailing was much less fun, however, because the only way to steer was with the paddle. I soon made a plug for the trunk, and went back to boardless sailing. A board well aft—behind the backrest, perhaps—might be worth trying sometime.

The *Wailer* was built with resorcinol glue, and has been fine. *Bob* was built with plastic-resin glue, and lost some of its shape as soon as the mold came out. It hasn't opened up, however. Epoxy is the glue I use most often now, ignoring the hoopla surrounding the various brands. Resorcinol and plastic resin have longer pot lives, and take much of the tension out of the work; but no waterproof glue is pleasant to use, or good for your health. I use rubber gloves when applying it, and a respirator when sanding it. You had better too.

Dread Bob and the *Kingston Wailer* have served us well for eight years now. They are not finicky or fragile, and can be bumped around and run up on beach or marsh, and loaded down with gear, tools, or salvage. When on land, they probably shouldn't be stepped in or sat upon. They are reasonably dry in a chop, but would need much more extensive decks for open water. They are flat-water boats, but the lack of decks makes them lighter, easier to get around in and store things in, and more pleasant to be in. We use them all the time.

J. Henry Rushton

The surge of interest in canoeing after the Civil War may have been prompted by the pacification of the wilderness: white men could play there, now that red men and animals were gone. This lasted less than 50 years before the boats became increasingly complex, perfect, and expensive. According to Mystic Seaport, a top-quality 16-foot sailing kayak cost as much in real money in 1903 as a modest new car costs today. By 1903, technological improvements to the bicycle were drawing people away from the water and onto the highway.

The canoe craze really started in England. In 1866, John MacGregor published his instantly successful *A Thousand Miles in the Rob Roy Canoe*. MacGregor was a friend and fellow religious fanatic of Robert Baden-Powell, founder of the Boy Scouts. The idea of the Boy Scouts—and Baden-Powell frankly said so—was to keep boys too busy and make them too tired to masturbate. Most likely, MacGregor shared this laudable aim, but so strong was the impetus he gave to canoeing that soon even married people were taking up the sport.

The distinction between canoes and kayaks has never been perfectly clear. Is it the paddle, the decking, or the posture of the paddler? To me, the paddle makes the difference, and that is what the organizers of the Tuckahoe River Race felt this year, when they decided that an entrant in a 13-foot Old Town pack canoe, sitting up on a thwart, should still be in the kayak class, because he swung a double-bladed paddle.

J. Henry Rushton, America's most famous builder, used the word canoe exclusively. He built wonderful kayaks just the same. Rushton was a small, frail man from upper New York State. At first he worked in some of his father's various enterprises—a sawmill, a general store, a hotel—and at 31 built himself a boat to enjoy the outdoors. A round-bottomed clinker rowboat with steamed frames, it seems a remarkable first effort; although before the days of advertising and television, people had less ephemeral junk in their heads. They were handier and more versatile.

Even then, Rushton's boat stood out. When another man wanted it, he couldn't resist selling, and so it happened a number of times. He sent two boats to the Centennial Exposition in 1876, and the next year he moved out of rented quarters and into a new shop designed for boatbuilding. By 1881, he had outgrown that shop, and built a new three-story one where he kept 20 to 30 workers busy until he died in 1906.

The biggest difference between Rushton's kayaks and the English ones like *Rob Roy* was that they were planked with white cedar, not oak, and so weighed less than half. If the object of kayaking is to tire the paddler, then the heavier the better, and especially if the trip includes portages. But from his own experience in the Adirondacks and from talks with other outdoorsmen, Rushton concluded that most paddlers wanted a light boat. He touted the weight of his boats, and constantly experimented to make them lighter still. The lighter a boat is, the better it must be built if it is to hold together.

Rushton found a valuable publicist in George Washington Sears, a woodsman even frailer than he. Sears, who wrote books and articles under the pen name Nessmuk, wanted a kayak weighing less than 20 pounds. Over the next few years, Rushton built him five, starting with *Wood Drake* (17½ pounds, including 2 pounds of paint), and culminating in *Sairy Gamp* (10½ pounds, with precious little paint). These Nessmuk canoes were 9 or 10 feet long: too short to go easily through the water. But they were certainly long enough and heavy enough for a tubercular old man to lug through the woods between lakes.

Commercially-produced Nessmuk models were a bit larger and heavier: "Length 10½ feet," says the 1903 catalog, "beam 27 inches, depth amidships 9 inches [meaning 4- to 5-inch freeboard], weight about 18 to 22 pounds." Planking was 3/16-inch white cedar. Steamed frames, on 1- to 3-inch centers, were half-round in section to save the weight of rectangular ones. Outwales and inwales were spruce. It seems impossible that these boats, without spreaders or thwarts, should have held up and held their shape. I doubt that they did for more than a few seasons, because from the time wood is fully dry it begins slowly to lose its strength—its elasticity and ability to hold fastenings—and these boats have no reserve strength anywhere. That they lasted at all is a tribute to the virgin forests where Rushton's timber was cut, and to his wonderful designs and workmanship.

The quality of materials and work that went into a Rushton kayak is hardly possible today, even by a hobbyist. In Rushton's time, such work was done by paid hands, and they weren't marvelously paid either. The boats were clinker-built, but not in the usual manner. The laps were cut at 45 degrees, so they fitted flush outside and in (Figure 1–8). They were held together by copper tacks driven through and clenched over, spaced close and alternating, one driven from outside and the next from inside. To do this, planks at the turn of the bilge were spiled on thicker stock, hollowed inside, and rounded outside. Talk about work! The seams were luted with shellac, and not because Rushton disliked polyurethane sealants.

Many of Rushton's kayaks were decked and could be sailed as well as paddled, using vast, inefficient bat-wing ketch rigs. He developed a line of kayak hardware, or "canoe jewelry," as he called it: nickel-plated mast partners with three tiny cheek blocks each side, kick-up rudders, fan-like centerboards. This stuff was all light and beautiful, and it worked. Kayak building would be more interesting (and more expensive) if it were available today.

Clinker Kayak

I built a Rushton-type kayak for the experience of doing it. I felt no temptation to improve on his lines, but with a reprint of his catalog before me, selected what seemed a good size and shape for my 160 pounds. Some of his boats show a hollow waterline, and Rushton says that they are easier paddling. But a hollow waterline always looks like a sorry attempt to compensate for a too-fat boat, so I chose one with a straight waterline. The bow half-angle of my boat is 12 degrees, compared with 16 degrees

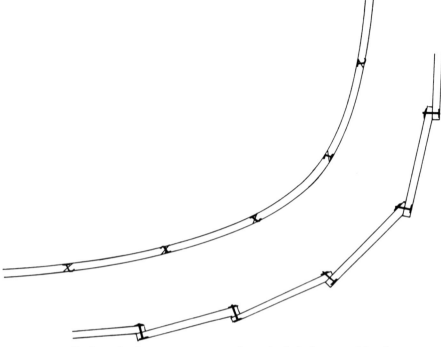

Figure 1–8. Clinker planking: conventional method (below) and Rushton method (above).

on *Dread Bob,* and because the keel has no rocker and the ends are deep, no skeg is needed to keep her on course. My boat has no tumblehome, which never serves but two purposes: for looks, and for a nuisance in the building process.

My boat is 14 feet by 28 inches, and 14 feet by 25 inches on the waterline. Being round and no beamier than *Bob,* she, of course, has less stability. Midships freeboard is 6½ inches, and is carried perfectly straight through two-thirds of the overall length, curving up 2½ inches at bow and stern. The intention was the minimal sheer that people expect to see, with resultant minimal windage. Mass producers must cater to a va-

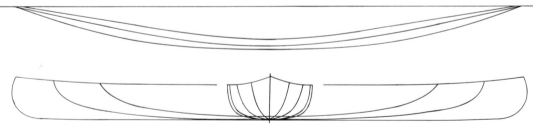

Clinker Kayak. Lines to inside of plank.

riety of buyers, some of whom want to play Injun; hence the exaggerated sheer on many boats, as well as birch-bark patterns printed on plastic topsides.

The boat was built upside down over seven molds of ¹/₂-inch plywood. A keel of ³/₄-inch by 2-inch mahogany was rabbeted to receive the planking, and notched into stem and stern of ³/₄-inch plywood. Two bits of ³/₈-inch plywood (*pace*, Henry!) made permanent frames 9 inches from each end. This plywood is all hidden under little decks, which are, of course, white cedar.

The planks were spiled onto ³/₄-inch cedar flitches, and were then re-sawn and dressed to make a ¹/₄-inch plank for each side. Clinker planking is fun and not difficult, but every single move must be thought out before the first one is made. Carvel planking forgives your errors; clinker plank-

Figure 1–9. Clinker kayak stem and bulkhead.

ing magnifies them. The width of every plank at every mold and the placement of every fastening (especially the fastenings in the bent frames that aren't there yet) must be constantly kept in mind. Nevertheless, a clear-headed builder can hang clinker plank pretty quickly, and it's nice to finish each one before hanging it, and be done with it. Nice to leave fastenings showing, too, instead of plugging and sanding. All boatbuilding saves too much finishing work for the end of the project, and clinker gets some of it out of the way as you proceed.

The boat has six planks to a side. I screwed them to the keelson, the stem and stern posts, and to each mold. Frames were to be on 5½-inch centers, and between where each pair of frames would later go, I fastened the plank laps with copper clench nails. If the planks had been thicker, I'd have used a bronze screw, and not had the tail of the nail showing inside. Rivets and roves are not handy for a builder working alone, because they need three hands to draw up. In every lap, I ran a thin bead of polyurethane sealant, just like Rushton would have done.

My friend Dick Shew builds cedar whitehalls in Maine, and he says, "I'm not looking for anything special when I buy cedar." So often, he has spiled out a plank on the best stock, resawn it, and found it not so good inside. Now he uses run-of-the-mill stock, and a variety of jigs and clamps to cut away bad sections and scarf in others. Dick would like to use the quality of stock that Rushton did, but it isn't available, and he's tired of the shams. Today's cedar is second growth, which is the only kind of wood the world will soon have. Unlike hardwoods, cedar may be a sustainable harvest: My own favorite sawyer is now cutting over swamps that his father cut 50 years ago. It's still an excellent boatbuilding mate-

Figure 1–10. Four planks over molds and keel.

rial, but it handles differently from virgin forest stock, and we'd better get used to it.

The wood is less dense, because the tree has grown quicker, having more light and room. Therefore it is lighter in weight, and we can afford to use thicker planks. In fact, we *must* use thicker planks, because it's also less strong. None of this is damaging, but second growth also dents and nicks easier, and that's a pity.

It has more and larger knots, because more branches grow and live longer, so it must be cut to more waste than Rushton had. Sapwood is also proportionately wider, again making more waste. The brown knots are not troublesome, and can be painted over. If you are fastidious, you can give them a dab of epoxy first. Black knots are loose, or soon will be, and must be removed and plugged. Knots in the lapping edges of planks should be avoided when possible; and knots in the gains, where the planks taper out to nothing at the ends, should be avoided at any cost.

At the gains, I use a dory lap instead of a rabbeted lap, judging it less likely to split out later. On the kayak, the planks ran wild of the posts, and were cut off later in pairs, flush with the razor edge of the plywood. Their hood ends were later covered with a piece of steamed sassafras, again bedded in polyurethane.

Steam Bending

Sassafras is a wonderful steaming wood, and oak a much overrated one. Rushton used red elm, but most elms are gone now. Sassafras weighs three-quarters what oak does, and holds fastenings nearly as well. It has far more decay resistance than oak, especially as there is only one species of sassafras, and your sawyer has less chance to bamboozle you. When steamed, oak grows a crop of raised grains and furry splinters, and must be resanded after bending. Sassafras dries smooth after steaming, ready for paint. Whether sawn or steamed, it gives off a lovely root-beer smell. Here in New Jersey, the trees are approaching the northern limit of their range, and are much deviled by bugs and fungus. The sawyer likes to sell the bad along with the good, so the flitches must be studied carefully before buying.

Green wood is not the only kind that can be steamed. Wood cut several years ago does just as well, but the moisture takes some time to come out of the wood, and so will take some time to go back. Best to cut and plane the stock a week or two before it's needed, and let it soak in water, weighted down. Then it steams like a green stick. The purpose of the steam is not to moisturize the wood, but to heat it without drying it out.

Most boatbuilding is fun, but bending hot frames is magic. When I have a batch to do, Carol takes a day off from work to get in on the fun.

With the molds still in place in the kayak, every possible rib was bent in and fastened. The outwale was screwed on. The molds were removed, and the final ribs were bent into their places and fastened through the holes that had held them. The inwale was screwed through the ribs and planking and into the outwale. The decks and backrest completed the job.

Outwales and inwales were $\frac{1}{2} \times \frac{3}{4}$ inch, from a good stick of dark red lauan. Such sticks are rarely found in today's local lumber yards, though they were routinely hacked up for ranch trim 30 years ago. Lauan is increasingly closely graded, and local yards usually buy "slightly wormy" grade, which usually breaks at the worm holes when bent. Kiln drying has fixed the worms but not their holes, which run cross grain and will easily pass water through a 2-inch plank. The boatbuilder needs to look farther for his lauan, perhaps to a specialty yard. Douglas fir, which comes from our own virgin rain forests (just as lauan comes from the virgin rain forests of the Philippines) is also graded more and more closely, and the faults allowed in each grade multiply, so the buyer must now select with increasing care. I never buy these woods without thinking what I'm helping do to the planet. These problems are beyond the scope of a book about boats, but we should all buy and use materials sparingly, and build boats no larger or fancier than they need be.

The clinker kayak is a much handsomer boat than *Dread Bob*, and somewhat better paddling. In a recent race, the clinker boat with flat-bladed paddle and *Bob* with curved paddle were neck and neck after three miles. It's surprising the clinker boat isn't faster still, with its sleeker lines and no skeg to drag. The problem is the extra wetted surface of the laps. If the planks are $3\frac{1}{2}$ inches wide and $\frac{1}{4}$ inch thick, then the clinker

Figure 1–11. The clinker kayak.

boat has 8 percent more wetted surface than a boat the same shape built by Rushton's method. It matters. Years ago, when people didn't understand what caused resistance, they spent a lot of time looking at lines drawings and fingering half models. Today, hulls are more likely to be described with numbers—measurements and ratios. In a 14-foot kayak with an all-up weight of 200 pounds, the resistance in pounds looks like this:

KNOTS	FRICTIONAL	WAVE-MAKING	TOTAL
3	0.8	0.2	1.0
4	1.1	0.8	1.9
5	1.6	3.0	4.6

Even wave-making resistance, the last defense of people who like to judge hulls by eye and feel, can largely be spelled out with numbers and proportions: length, length-to-beam ratio, displacement, bow half-angle. To find the horsepower needed at the various speeds, multiply total pounds resistance by knots by .003.

My clinker kayak weighs 40 pounds—not much more than Rushton's standard kayaks that size. Under sail, she will not steer at all without a paddle. She will take plenty of hard knocks with no worse result than a dent in the planking. She is comfortable to sit in, and has never leaked. Like all clinker boats, she is noisy through the water. Almost any size paddler can find one rib or another that will do for a foot brace. Wherever we take her she stops crowds, and that's some reward for 103 hours of labor.

The DK 16

In 1979, when the Gougeon Brothers published their book *The Gougeon Brothers on Boat Construction,* they regretted that, to their knowledge, four daysailers were the only tortured plywood boats for which stock plans could be bought. Wistfully they said, "Canoes and rowing shells are another potential area." Ten years earlier, a British shop teacher had had the same thought, and had developed a series of kayaks for his students to build. Still, news of Denis Davis's work did not reach American builders until *WoodenBoat and Boatbuilder* published his articles about the first of them, the DK 13, in 1986. Though perfectly round bottomed, the boat was so simple that complete plans could be published in the magazines.

At the time, Carol was wanting a faster boat than the *Wailer* for the Tuckahoe River Race. The DK 13 seemed a big step up in sleekness, but in studying the drawings, I wondered why it had to be only 13 feet 8 inches overall. For turning quickly, that's the right length, and some keel rocker is useful; but for flat-water racing, the keel line should be perfectly

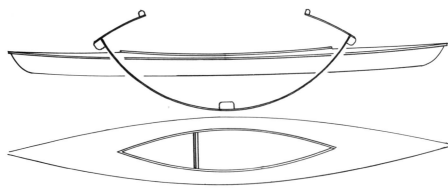

DK 16 Kayak.

straight, and the boat should be as long as two scarfed sheets of plywood could make it. I fiddled Davis's offsets. Later, *WoodenBoat* published another article about the DK 16, and Carol's boat is very much like it. As Davis promised, she went together quick and easy.

After the two bottom panels are scarfed up and cut out to a series of offsets shown on the plans—like the follow-the-dots games children play—their bottom edges are glued and screwed to a keelson about half the length of the boat. The forward and aft ends are then wired together, and the resulting hull shape is as magical as anything a child achieves when the last dot is connected.

Figure 1–12. The DK 16—a curved hull from flat plywood.

Usual practice is to wire with copper, bury the wire in muck on the inside, and glass tape over the whole mess. The outside wire is then cut off, ground, and glassed. I prefer to wire with baling wire, tape *between* the wires inside, and draw them before taping the outside. Inside, epoxy putty around the wire holes makes a keel or chine plenty strong enough for a small boat. The work is lighter, quicker, and neater, and taped-seam construction is lumpy enough at best.

Davis used an inwale, and to it he attached a very thorough deck, which could be fitted with a fabric skirt by the Greenland roll enthusiasts. Nothing makes us as nervous as being tied into a boat that can capsize. If over she goes, even Carol who can't swim would rather be in the water by herself, with all four limbs at her service. I gave her boat an outwale, which widens it but is too close to the water to interfere with the stroke. A highly cambered deck was necessary to stiffen the hull and raise the freeboard near the paddler's waist, but I left a deck opening big enough to lighten the boat as well as to set the paddler at ease. Varnished, she weighed just 19 pounds, a weight even Rushton couldn't have matched in a half-decked kayak 16 feet by 27 inches. Her shape is in no way inferior to his designs. The secret can be told in one word: plywood.

The DK 16 is squirrelly to board. The planking isn't strong enough to step on, so you must step on the keelson, and get your bum down on the cushion as quickly as possible. Even seated, the beginner is likely to say, "Whoa! Whoa!" and flail the paddle around for a minute. The bottom is the arc of a circle, and waterline beam is less than 19 inches. But she picks up buoyancy when she heels, and if you don't panic and push down

Figure 1–13. The DK 16 is a very sleek boat.

on the low side, she'll come back. You get used to her, and you love the way she moves with the lightest stroke of the paddle.

With a 160-pound paddler aboard, she has 17.4 square feet of wetted surface. With the same load, the *Wailer* has 18.7, *Bob* has 22.4, and the clinker kayak has 23.9. With a bow half-angle of 9 degrees and a length-to-beam ratio of 9.3, she's not a big wake-maker, and that's especially important for a racer, operating most often at the upper end of the speed range. The DK 16 tracks reasonably straight. Her worst failing is fragility, and she's not to be run into snags or up on the beach very often. Slather her with fiberglass (as I've seen done on other DK 16s) and the weight builds up much faster than the strength. Fiberglass also requires finishing of an especially noxious kind, and one virtue of plywood is that it comes factory finished.

For a couple of years, Carol and her DK 16 did very well in the race. Once they beat Dick Young, who is practically a Tour de France bicyclist and doesn't even smoke. But at 45, she is finding less adrenaline for this kamikaze effort. Next year she expects to take the *Wailer* again, and let Dick have the DK 16. None of our kayaks has ever won the class, and it's time one did.

Building a tortured-ply boat from stock plans, you may get a good boat, but you won't learn much about boats or boatbuilding. The same is true of taped-seam boats, whether tacked or wired together. If the plans show each measurement and each dot to be connected, and if you study them long enough, you'll realize that the designer has had all the fun and left you the work. On the other hand, designing your own tortured-ply boat is the most exciting carpentry I know, and the most educational.

Tortured Plywood Rowboat

Guidance on designing tortured-plywood hulls is not plentiful. The Gougeons confine their discussion to multihulls, and narrow ones at that. Sometimes you see an old Moth—an 11-foot sailing dinghy—and realize that at one time builders were working hard to learn how plywood could be tortured to suit a planing shape. What they learned is not accessible today.

Anyone working with sheet goods (e.g., plywood, metal, cardboard) soon learns that if a sheet is twisted, it forms a compound curve. This curve will be the surface of a cone, and can be predicted on a drafting board, as explained in Chapter 3. However, sheet goods can also be tortured to form compound curves that are not conical surfaces. What these curves will be is much harder to predict. The drafting board is useful only to show the shape you *hope* to achieve. A model will tell whether there is

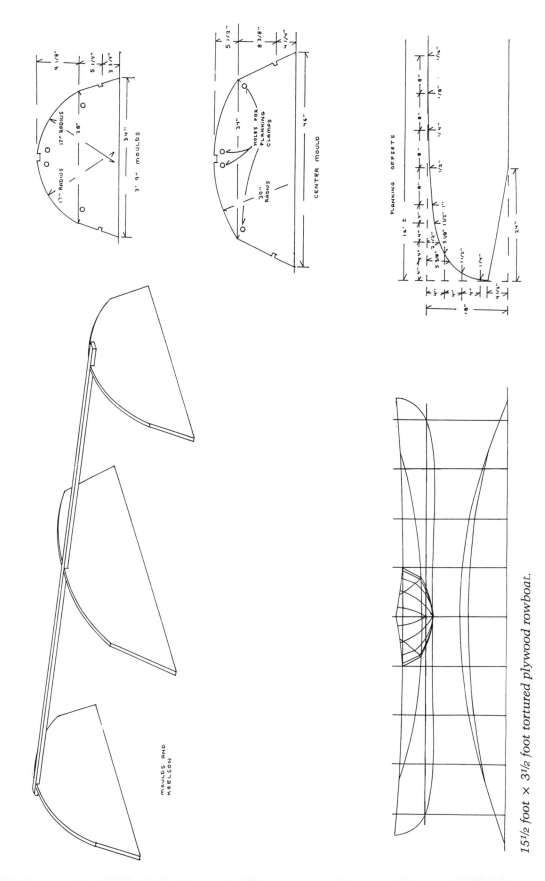

MOULDS AND KEELSON

3' 9" MOULDS

9 1/8"
5 1/4"
3 3/4"
17" RADIUS
17" RADIUS
28"
34"

CENTER MOULD

5 1/2"
8 3/8"
4 1/4"
34"
HOLES FOR PLANKING CLAMPS
30" RADIUS
46"

PLANKING OFFSETS

16' Z
1/16"
8"
1/8"
8"
1/4"
8"
1/2"
8"
1"
4"
1 1/2"
5 3/8"
2 1/2"
5 1/8"
1 1/2"
1/4"
24"
4"
4"
4"
1 1/2"
8"

15½ foot × 3½ foot tortured plywood rowboat.

any *possibility* of achieving that shape. But only building the hull will tell whether it can really be done.

I am not fond of rowing. It is certainly an efficient use of the body's limited energy; but not seeing where I'm going seems to annoy me more (and get me in more trouble) than most rowers. I began the tortured-ply rowboat because I was excited by the DK 16 and wanted to try a more original project. A fixed-seat boat with foot brace and enough beam to mount the oarlocks on the gunwales seemed sporty enough for a reluctant oarsman. Outriggers are awkward around docks, and they and sliding seats are difficult to make. Units can be bought and dropped in, but they weigh at least 20 pounds, and they cost more than I planned to spend on the entire boat. Carol isn't keen on rowing either, so once testing was over, we doubted the boat would be used often.

Racing shells get away with very narrow beam, as the oars play a big part in keeping them from capsizing. The center of gravity of a rowboat is higher than a kayak, because the crew sits higher, and the boat must have more freeboard. Oarlocks must be 8 inches above the seat, or your hands will hit your knees on the recovery stroke. So it seems a rowboat should have more waterline beam than a kayak. I figured a 25-inch waterline beam, like *Dread Bob*'s, would make a rowboat about as tippy as the DK 16, and that proved to be a good guess. If waterline length could be 15 feet, that gave a sporty 7.2 length-to-beam ratio. Eighth-inch plywood would easily take a curve of 30-inch radius (it's 16-inch radius in the DK 16), and this would give the necessary displacement with 25-inch beam and 3-inch draft. Continuing this arc up to the gunwale 11 inches above waterline gave a somewhat excessive 51-inch overall beam, but it seemed a possible starting point, especially for a designer who wanted to get away from the drafting table and on with the woodwork.

Here's where the fun begins. In most kinds of boatbuilding, the creative part of the work is done on the drafting board, and the building is mere craftsmanship, however challenging and rewarding that may be. But in the earliest boats of nearly all cultures—birch bark and dugout canoes, Nile riverboats, Viking longships—the creative part was the building itself, because the skin was made first, and any framing was added later. Today, a few builders of clinker boats still work this way: The master hews and rabbets a keel, sets up stem and stern on it, and begins planking without molds. He shapes the planks by eye, by feel, and by experience. He knows what changes in their curved edges will make a boat narrower or wider, fuller or finer in the ends, and what will produce more or less deadrise or a harder or slacker bilge. He designs the boat *while* building it, and the work must be glorious—more like a sculptor's work than a craftsman's.

Such clinker building requires a long apprenticeship, and judging from the boats produced that way, it doesn't encourage innovation. You know what works, and with minor variations you stick to it. With tortured plywood, the excitement of designing while building is just as intense, but by working with models first, the most obvious errors can be spotted without having to throw a whole boat away so there's more opportunity for fresh thought.

For modelmaking, the Gougeons use aircraft plywood. If the model will be one inch to the foot, they recommend that the skin thickness be scaled down less: $1/4$-inch-ply boat, $1/16$-inch model; $1/8$-inch boat, $1/32$-inch model. Such plywood is available, but I used $1/32$-inch balsa from the hobby shop, and it worked fine. I built the model on a strongback, using white glue and holding the parts together with straight pins until the glue dried. These materials can also be used for tow-test models, if a couple of coats of oil paint are added to protect the glue. White glue is not toxic like epoxy. It is easier to use and clean up, sets quicker, and may be just as strong. Some day we may have a waterproof white glue.

The model quickly revealed that sheet goods couldn't be tortured to the shape I wanted. What happened—what usually happens when you try to bend sheets to a shape they won't take—is that they took a concave curve, not a convex one. The problem area was where it often is in tortured ply, in the topsides amidships. Reluctantly, I realized that I must have a chine, though it could be well above waterline. With a chine, maximum beam could be held to 3 feet 6 inches, and the oars could be shorter.

The full-size boat was made by scarfing up two $1/8$-inch plywood panels, 18 inches by about 15 feet 11 inches. They were bent around a midships mold and two frames of $1/4$-inch ply. They were glued and screwed to a $3/4$-inch by $1 1/2$-inch keelson that notched into the mold and frames, and was glass taped to the frames. Cellophane tape was the parting agent on the mold. The centerline seam beyond the keelson was wired together and glass taped, as in the DK 16.

The resulting bottom shape was slightly deeper forward and aft than amidships: like a double-ended coble. Slightly different offsets on the keel line would have prevented this. I doubt that it increases the resistance, though it would have been anathema to builders a hundred years ago. It must slightly increase the prismatic coefficient also, which might be beneficial in the upper end of the speed range. The higher the speed, the higher the prismatic should be.

Eyeballing

A chine log $1/2$ inch $\times$ $3/4$ inch was slipped in behind the planking, in notches already cut for it in the mold and frames. It was glued in place, without worrying about fairness. If the edge of a piece of plywood is a

Figure 1–14. Tortured ply rowboat under construction.

straight line, it will always form a fair curve, no matter how the sheet is bent or twisted.

The chines rose beyond the sheer about 2 feet from the ends, and a piece of topsides planking was easily enough added to give more than adequate freeboard amidships. From calculations at the drafting board, I knew how much freeboard I wanted at bow and stern, and at the oarlocks. I clamped a light batten to the sheer at those three places, and with other clamps in my hand (spring clamps are good for this kind of job) began to walk around the boat, sighting the batten.

Eyeballing is a skill the boatbuilder must learn. John Yank taught me, and you will be lucky to find a teacher as good. You must learn to see, and to trust what you see. You are looking for flats, bumps, and hollows—just as you would in lofting—and trying to fair them out into a harmonious curve. In the old days, when people didn't know about wetted surface, wave-making, proportions, and the math that tank-testing has revealed, they thought that unfair curves in a hull meant high resistance. Today we know better. But fair curves are still what the eye yearns for in a boat, and deliberate or careless unfairness can often lead to building complications. So squat down and sight the curve in one plane. Stand on a chair and sight it in another. Move forward; move back. Eyeball it from as many angles as your shop allows, and make the little adjustments that are so little trouble and make so much difference before cutting or planing the wood for the last time.

Outwales and inwales, 3/4 by 3/4 inch, firmed up the sheer. Spruce is a good lightweight wood for such jobs, and I've had no trouble with Engelmann spruce on boats kept indoors. Like Western white pine, Engelmann has no rot resistance at all. For boats kept outdoors, Sitka spruce or New England white pine are the minimally durable lightweight woods.

The tortured-ply rowboat weighs 40 pounds with seat, paint, and oarlocks (I like the lollypop kind, that stay with the oar, not with the boat). If it weighed twice as much, total displacement would only increase 14 percent, because the oarsman is most of the load in any case. Speed varies roughly with the cube root of displacement, so if this boat has a top speed of 6 knots, and the same power was applied to the same boat weighing 80 pounds, only 1/4 knot would be lost. But that 1/4 knot is worth having; and on land, doubling the weight of a boat can make the difference between car-topping and trailering, between handling it yourself and dragging a helper along.

On the water, the tortured-ply rowboat is as tender as predicted, and some "whoa, whoa" is usually said until the oarsman is seated and his heartbeat slows down. Like the DK 16, the rowboat finds reserve buoyancy in heeling, and it has never capsized, though it is only used a few times a year. A pull on the oars gives a gratifying surge forward (backward, really, from the oarsman's vantage). As a non-rower, I find it runs perfectly straight with the maximum sustained power I'm able to give it, and is low enough to be little blown about in cross winds. If I try to sprint with it, steering becomes more demanding: the shore astern must be studied, and extra power applied to one oar or another to keep it on course.

Keeping a rowboat on course gets back to the essential problem of rowing: You can't see where you're going. Periodically, someone will market an oar mechanism that reverses your power, so that you pull backward but go forward. I haven't seen these contraptions, but in pictures they look cumbersome, heavy, and expensive. And some power loss seems inevitable. In the days when people rowed for a living—fishing, hauling, bum-boating—they often rowed backward, and did see where they were going. To row this way for long, with efficient use of your muscles, you ought to stand up, and that requires a boat more burdensome than would be acceptable for recreation.

I tried a bicyclist's mirror, clipped to my glasses frame, and that nearly drove me nuts. A bicyclist knows which side of the road he's on. He looks in the mirror and sees the bow or stern of a car, and has no doubt whether it's coming or going. But that stump or that bridge abutment gives the oarsman no such message. It may be to port or it may be to starboard—

Low-Resistance Boats

Figure 1–15. Tortured plywood rowboat in the Tuckahoe River.

wherever they are—but be sure that the mirror will tell you the reverse. And a rowboat with an ⅛-inch plywood skin doesn't win many battles with stumps or bridges.

Complete plans for this boat are not given. A designer so little interested in rowing has no business laying down the law about rowboats. Some more enthusiastic oarsman may want to use this plan as a starting

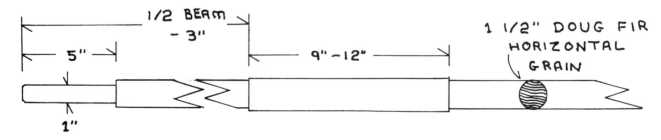

Figure 1–16. 7-foot spoon oar.

point for his own tortured-ply experiments. If so, I'd enjoy hearing the results.

We use 7-foot spoon-blade oars in this boat. Shafts are fir, and must be larger in diameter if of spruce or pine. For its weight, Douglas fir is the strongest of all woods, and first choice for all structural members—if you can put up with its tendency to check and spall. It's hard to distinguish fir heartwood from sapwood, but that matters little for oars that are kept under cover. Ideally, there shouldn't be sapwood under the leathers.

The oars are made much like the kayak paddle, but with different blade shape, copying the shape shells use. Weight saved outboard is very valuable, and top and bottom sides of the shafts can profitably be pared down by those who don't mind busting a pair to see how far they can take it. Inboard, shafts are often left square, to help balance the oar in the lock. The Portuguese, who swing very long oars, often bolt on an extra weight inboard. Oars are too much trouble to make, or too expensive to buy, not to be leathered. Buttons (the collars at the top of the leathers) are essential the first time you ever row, and a damned nuisance thereafter. I agree with Phil Bolger that oar handles should be 1-inch cylinders, with no "foolish swell."

In fact, it is probably foolish to say anything about fixed-seat rowing, because Bolger has said it all. He is quite possibly the best writer who ever wrote about boats, and the rowing boat is his favorite. If he rows for exercise, he doesn't say so. He does it for pleasure, to be on the water, to

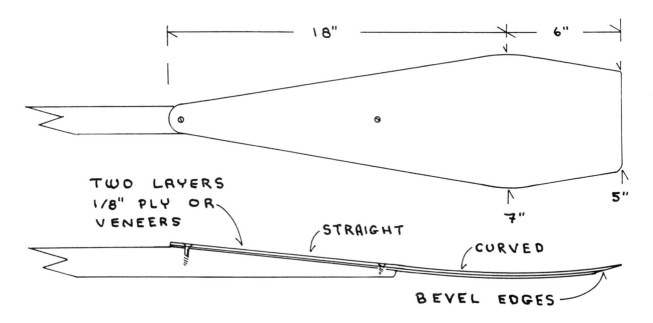

TWO LAYERS
1/8" PLY OR
VENEERS

STRAIGHT

CURVED

18"

6"

7"

5"

BEVEL EDGES

see the scenery, and to muse. Therefore, I hesitate to take issue with him, but I do like the 7-footers in this boat, and he'd have them much shorter. His formula (32 percent of the oar length inboard of the lock) would give toy-like oars 5½ feet long. In the *Dobler* (see Chapter 3) with 5-foot 2-inch beam, the 8-foot spoons exactly fit Bolger's formula, and seemed exactly right. I suspect that as the oarsman remains the same size, no matter what the beam of the boat, oar length must take his size into account, and oars for wide boats will be proportionately shorter than oars for narrow ones.

2. Garveys

Bolger likes flat-bottomed motorboats, and he likes them plumb-sided. But he's found that plans are easier to sell if the boats are flare-sided, because then people call them dories. When it comes to motorboats that are square on both ends, people like to call them garveys, not scows or prams. A garvey is nothing more than a scow with a bow that is round in profile—a spoon bow. Many people think making that spoon bow is pretty tough, so as often as not they make it flat, and call the boat a garvey anyway.

The 1953 edition of *How to Build 20 Boats* published by *Mechanix Illustrated* contains many curiosities. There is a 175-pound strip-planked canoe, a houseboat in which every line but the portlight rims is straight and either plumb or level, a cruising flatiron with a concrete keel by William Garden, and a flat-bowed garvey by Howard Chapelle. What these boats have in common is the determination to avoid some step in the usual building process—a step that was thought too difficult for amateur builders—like steaming frames or cutting bevels or casting lead. They are like arc-bottomed kayaks with flat bottoms. As a result, they have not lasted. Thirty-eight years later, well within the life of a wooden boat, you never see a one of them, because when people are looking for an old boat to fix up and keep going another few years, they pass over the 175-pound canoes and their ilk.

When garveys were built for work, the spoon bow was not thought too difficult. Topsides planks were cleated together on the floor and sprung around a mold or two. Often the topsides flare gave about the right rocker for the bottom, as it does in *Bob* and the *Wailer*. The bow curve was cut, and a chine log clobbered on. At the bow, the log was pieced as necessary to take the curve. The cross-planked bottom was whacked down, the boat was flipped over, and thwarts were stuck in to hold the shape before the molds were pulled out. This, and some careful eyeballing, is literally all there was to it.

Russ Adams Garvey

The garveys Russ Adams built for many years on a semi-production basis at his small shop in Somers Point, New Jersey, were not much more complicated. His molds looked like saw horses. He had to shape the bottom

33

plank, because he built planing boats and wanted no rocker. He used better fastenings and more careful fits—very good fits, in fact, to judge from the Adams garveys still around 20 to 40 years later. These boats are still in demand along the Jersey Coast, and change hands for fair prices. But the flare and curve he put into the topsides gave the whole shape of the boat, except for the bow, which he cut.

Adams garveys are similar to other garveys built at the time, intended for fishing and clam tonging and working around marinas. They have the same plainness, the same cedar planking and lauan framing in off-the-shelf thicknesses, and they probably took no longer to build. So many of them are still in use because Adams put no extra weight into them, and thought carefully about the details, taking the trouble to get the proportions right and the lines fair.

Old Ralph Clayton bought a boat from Russ, and liked to hang out at the shop. He watched many a garvey come off the molds, eyeballing it and talking it over. Time came when Russ was tired of it, and glad that he had no orders, and could just sit and gam with Ralph. Then one day a stranger came in the door. "How much do you charge for one of your garveys?" he said.

"I used to charge $250, but I don't make them any more," said Russ.

"Well, how much would you charge if you did make another one?"

"$450," said Russ.

The stranger paid the $450, and left. "Damn it!" said Russ to Ralph, "I *still* didn't want to make another one!"

The plans show a boat adapted from the Adams garvey for plywood construction. When Adams was building, owners of boats this size expected to keep them in the water, but today most keep them on trailers. Marina space has gone up faster than inflation, and people want more things today, so they need to spend a little less for each one. An Adams boat didn't soak up much water, even after a winter on land; but today, not all builders are as skilled as Adams was, and good-quality materials are much harder to find. The plywood boat should be lighter, and if kept in the water should soak up less than cedar planking, and so be lighter still.

The big change in building technique is that the bottom determines the boat's shape, not the topsides. The bottom is built first, cutting out two sheets of $1/2$-inch ply 5 feet 3 inches long. Chine logs and then butt blocks are glued and screwed on. The bevel of the chine and transom logs is 15 degrees, and chine logs only go forward to station 0. At this point, set the bottom right-side up on sawhorses, making sure the horse tops are at least parallel to each other, if not level. Cut out the topsides and glue and screw them on. Make up and glue in the two frames and the double-

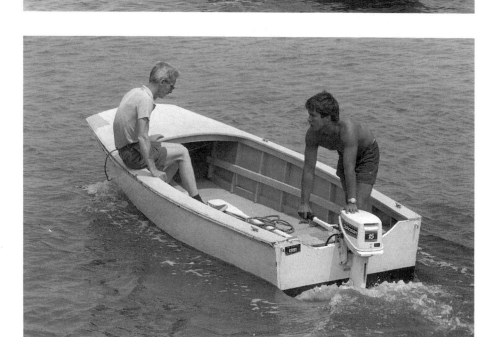

Figure 2–1. The Adams garvey shows a good turn of speed with modest power, and has a lot of obstruction-free working space.

thick transom. Pull the topsides together at the bow with a Spanish windlass, until they are 2 feet 8 inches apart at the sheer. The double-layer bow is made from a single sheet of ¼-inch ply, with the face grain running athwartships. Each layer is screwed to the butt block at station 0, bent around the bow, scribed, and cut to shape.

Assembling the bow will take a little time, and it's best to use a slow-setting epoxy and start the job when it isn't too hot. A dry run should be made first, and holes for wires drilled in bottom and topsides. By tapping on the outer layer of 1/4-inch, it's possible to tell roughly where nuts and bolts will be needed to draw the two layers together. The bow is then disassembled, and the parts and supplies laid out on a bench close at hand. The first layer is glued and screwed to the butt block. The faying faces of the two layers are thickly buttered with a spackling knife or a cement-finishing trowel. Screw to the butt block, pull down, and wire in place. Run the bolts in and tighten them up. Coated with wax or silicone, they will remove easily later. Taping the inside seams can be done now or later. Taping the outside must wait until tomorrow, after the bolt and windlass holes are being filled.

Sheer stringers, deck, guard rails, and bottom runners complete the boat. This will require four sheets of 1/2-inch ply, one of 3/8-inch, and one of 1/4-inch. Total weight should be 230 pounds, so this Adams garvey is

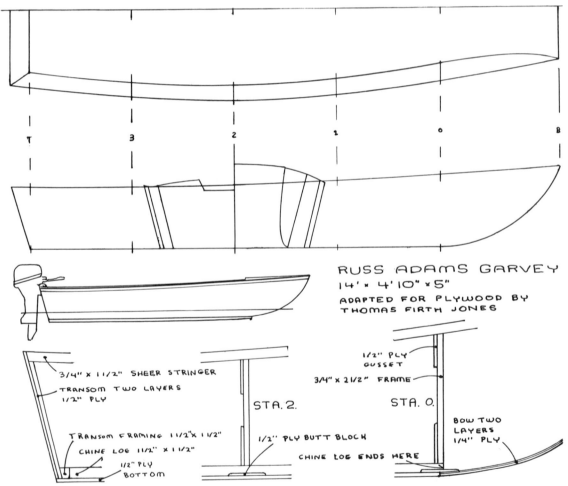

RUSS ADAMS GARVEY
14' × 4'10" × 5"
ADAPTED FOR PLYWOOD BY
THOMAS FIRTH JONES

3/4" × 1 1/2" SHEER STRINGER

TRANSOM TWO LAYERS
1/2" PLY

TRANSOM FRAMING 1 1/2" × 1 1/2"

CHINE LOG 1 1/2" × 1 1/2"

1/2" PLY
BOTTOM

STA. 2.

1/2" PLY BUTT BLOCK

CHINE LOG ENDS HERE

1/2" PLY
GUSSET

3/4" × 2 1/2" FRAME

STA. O.

BOW TWO
LAYERS
1/4" PLY

Russ Adams garvey adapted for plywood construction.

100 pounds lighter than a 14-foot Boston Whaler, and a better shape to boot.

Notch the transom to suit the motor you have, with the cavitation plate ½ inch to 1 inch below the transom bottom. When I tried Ralph Clayton's Adams garvey, she planed readily with two people aboard, and a very tired 15-horsepower outboard. This boat should do almost as well with 10 horsepower, and maximum safe power would be 25. This one might give 35 miles per hour, which is plenty fast enough to go on water.

Planing Powerboats

A planing hull is not like any other kind of boat. Some sailboats can plane off the wind, but what's usually called planing is just momentary surfing. To plane, a boat must lift itself out of the water by climbing over

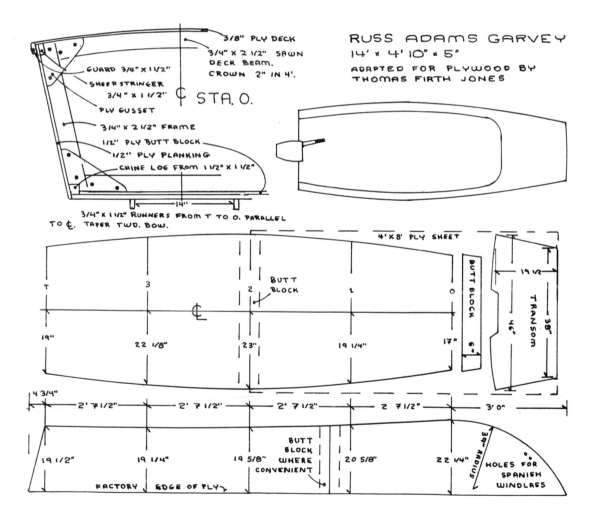

its own bow wave. To make this more difficult, the wave that formerly supported the transom is now astern of it, so the stern is down, the bow is up, and great gobs of power are needed. Sometimes you'll see the skipper of a planing boat push the throttle open, walk forward to get the bow down and the stern up, then walk aft and ease the throttle back. Once the boat is up on its bow wave, less power is needed to keep it there.

Ralph Clayton said that his garvey would run away from most bigger boats, and the claim is easy to believe. In planing boats, smaller is faster, especially as most boats with accommodations never quite make it onto a plane. They just stick their noses up and throw an evil wake, and everyone is sure he's having a hell of a ride. A true planing boat makes very little wake, because little of the hull is disturbing the water.

Steve, the college boy who worked summers at Ralph Clayton's marina, took me out to try the garvey. "This little boat is a lot of fun," he said. "It's so open, and you're so close to the water, you really get the sensation of speed." Indeed we did, though the speed couldn't have been 20 miles per hour. We stood up and walked all around in the big bottom, without worrying about trim. The drumming of ripples in the water was part of the fun. Bigger waves would be another story. Clayton tried the boat in the ocean once or twice, and said he didn't like it. Planing boats are not seaworthy, though some are worse than others. They pound at high speed, and at low speed, they broach.

The plywood Russ Adams garvey is a foot shorter than the original and four inches narrower, to suit 4 × 8 sheets. It still has plenty of room to stretch your legs, and you can do so with the boat at rest or on a plane. Thwarts can go anywhere, and Clayton had a thwart stringer the length of his boat, with enough fastening holes to suggest that he'd changed his mind more than once over the years. When I tried it, he had only one thwart near the stern. Best to go easy on thwarts, which aren't needed to stiffen the decked boat, and are often in the way. This boat is big enough for a couple of folding chairs. Some camping accommodations could be worked into it if kept light, except that it's fast enough to get anywhere and home in a day, so why not have a shower and sleep in a bed?

In a flat-bottomed boat like this, one horsepower will lift 50 pounds onto a plane. In a deep-V boat, one horse may lift half that much. Today, deep-V planing boats are popular with people who want a soft ride, to complement the rest of their soft, self-indulgent lives. They do sort of plane, with their snouts in the air and throwing a gigantic wake. Originally, the type was developed for offshore racing, and in those conditions it may survive a little longer than other bottom configurations. However, a recent *grand prix* event at Atlantic City made it clear that in a Force 7 (which would hardly trouble a displacement powerboat or a well-found sailboat) the racers couldn't take it. Some failed mechanically, some capsized, and some literally broke in half. Hardly any finished.

In sheltered water, the deep-V is uncivilized and unnecessary, and the tri-hull isn't much better. Planing hulls should have flat bottoms, or at worst the warped bottoms that are readily made from sheet goods, with a V-bottom forward and a flat run aft. And they should be light.

Paint

Paint this boat, and any other wood or fiberglass boat you build, with the best-quality semi-gloss latex house paint. Years ago, I asked a DuPont paint chemist why, if latex paint was so much better, DuPont still made oil. He said, "Because if we didn't, jerks like you would buy it somewhere else." Everyone knows that latex paint is easier and more pleasant to use. Sherwin-Williams says, "Oil paint has the best stain resistance and initial gloss, but latex is the longest-lasting finish, with the best color- and gloss-retention. It is least susceptible to chalking, and best resists peeling and blistering." They do sell both kinds.

The peeling and blistering of oil paint is caused by moisture in the wood trying to get out. Latex paint is porous, so it allows the wood to dry again after wetting. So does spar varnish (*not* polyurethane varnish), but varnish is best confined to the inside of a cabin boat, or a boat kept indoors, because sunlight eats it up. Sunlight isn't good for wood either, and the purpose of paint, besides decoration, is to shield it against sunlight.

You can't shield wood against moisture with oil paint. All you can do is retard the moisture getting out again. Epoxy resin is even worse, because it is less likely to peel and blister and let the moisture out. Epoxy resin is a terrible primer for paint, because it doesn't flow and even itself out like paint. Furthermore, epoxy hardens too quickly, and is hateful and unhealthy to work or sand. It does not penetrate wood any deeper than other surface coatings.

Epoxy resin does not prevent or retard rot. Rot is caused by several varieties of fungus, and their spores are everywhere. The fungus consumes sapwood almost before your eyes, but consumes heartwood much slower, especially if the species is rot-resistant. Pentachlorophenol (Woodlife) is a fungicide, but it is now illegal, like mercury bottom paint and lead in oil paint and other evil chemicals. Penta does not last in any case, but leaches out into the air, and must be reapplied. Linseed oil mixed with kerosene may have some ability to kill rot fungus. Cuprinol has none. Its purpose is to protect against teredo worms and other animals, not rot fungus. Pressure-treated yellow pine may look as if it contained copper, but the actual anti-fungus agent is arsenic. Another good thing to avoid.

To live, rot needs moisture, air, and the right temperature. In an open boat like the Adams garvey, air circulation keeps the temperature low enough so that rot doesn't thrive, and using good wood in the first place will guarantee a long life. A boat with enclosed spaces is more of a problem, and the best protection is generous ventilation to keep the temperature down and dry up moisture. If epoxy coatings work at all, they work by retarding the flow of air to the fungus. But ventilation works much better. And be sure that at least one side of every piece of wood in your boat—and preferably both sides—has a coating that can breathe. End grain, and especially plywood end grain, is another matter. Because the cell structure of the endgrain makes it suck up moisture faster than it expels it, an epoxy coating is a good idea here.

If you doubt me about rot, do not get your second opinion from people who hope to sell you their product. Read the *Wood Handbook* of the Department of Agriculture, or *Manual 250–336* of the Navy Bureau of Ships. They do not recommend the encapsulation of wood.

The hotter the climate where your wooden boat is kept, the shorter its life will be. With an enclosed boat, sunlight also makes heat, so a wooden boat will last longer in a cloudy climate than a sunny one. Wherever the boat is, ventilation and good materials are still the key. In any climate, horizontal surfaces should be painted white to reflect the sunlight.

A friend who spent much of his working life in yacht surveying also believes that southern species of wood brought north last well, and northern species brought south rot quickly. This may explain the poor reputation of Nova Scotia-built boats in southern waters. I have little experience here, but once did see a relatively new cruising sailboat—well built in New England—hauled in Florida after several lived-aboard years in the Bahamas. She was rotten beyond repair.

Wheelbarrow Garvey

For some years after Bob and Helen Diamond moved to their new house on Delaware Bay, simply looking out over the water was enough to satisfy their nautical ambitions. But then Helen took sailing lessons, and began talking about Hobie Cats. I suggested an Aqua Cat, which weighs half as much as a Hobie and has no jib or boom, so it's quicker to rig and kinder to less-experienced sailors.

The Aqua Cat was a success, and the Diamonds and their friends had many good sails on it. But it was still pretty heavy and awkward for two middle-aged people to drag up a soft beach, especially at half tide, and in practice they didn't use it unless they had company. Bob began talking about an easier boat, which the two of them (or perhaps just one of them)

Figure 2–2. Wheelbarrow garvey skimming the shore in Delaware Bay.

could launch and retrieve. He wanted to try fishing. Power could come from sail, or a small outboard, or perhaps just oars. What did I think?

I thought immediately of Chapelle's Seal Cove Skiff (*Boatbuilding*, p. 259), an 8-foot flatiron with a wheelbarrow wheel permanently fixed at the bow and handles sticking out past the transom. The alternative, a dolly or a removable wheel, would have to be carried in the boat or taken ashore after launching, and that could be a long distance on their very gradual beach. The wheel would have to make a big footprint, in any case.

Their part of the bay is always choppy, whether from wind or the wakes of sportfishermen. And the Diamonds are sociable people, with many friends and three grown children who visit often, so an 8-footer seemed small. But a bigger boat would be heavier, and the Seal Cove wheel was right in the bow, carrying only half the weight. The Diamond's boat would need a wheel further aft, like a modern wheelbarrow. This would keep the wheel out of the wave train, no matter what the trim. Even below the surface, the wheel would make considerable resistance, requiring many extra square feet of sail or a very determined oarsman. Outboard power seemed the best solution.

Compared to oars or sails, even a very small motor is enormously powerful. Bill Durham says that running in a 10-knot breeze, it takes a 1000-square-foot rig to develop one horsepower. But a 3-horse outboard is marvelously light and compact, and will push almost any 15-foot boat at hull speed, no matter what the shape or appendages.

We finally settled on an arc-bottomed, foam-sandwich garvey, 14½ feet by 5 feet. Square-ended boats give more buoyancy and carrying capacity than sharp-ended boats the same length or weight. The V-shaped bow and arc bottom are complications, but the bow justifies itself in chop, and the

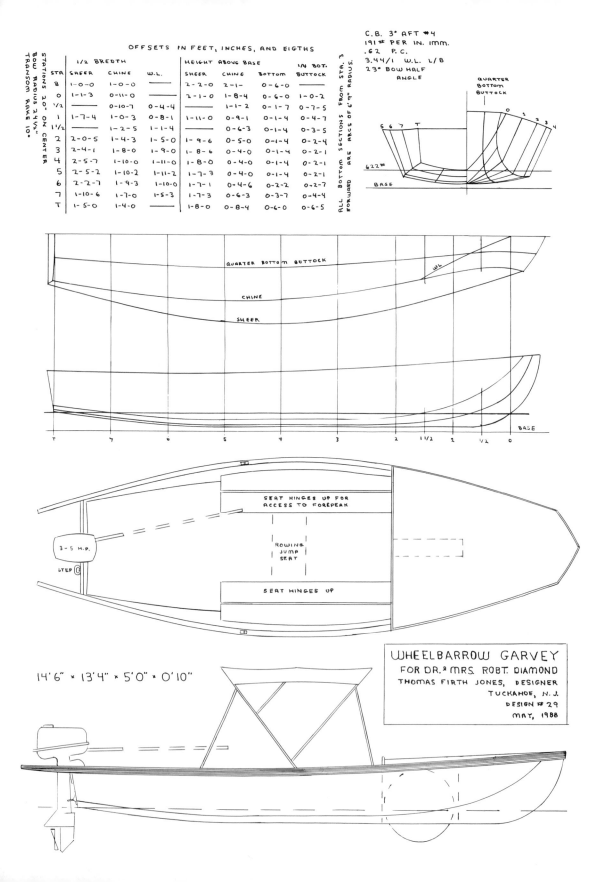

OFFSETS IN FEET, INCHES, AND EIGTHS

STA	1/2 BREDTH SHEER	CHINE	W.L.	HEIGHT ABOVE BASE SHEER	CHINE	BOTTOM	1/4 BOT. BUTTOCK
8	1-0-0	1-0-0		2-2-0	2-1-	0-6-0	
0	1-1-3	0-11-0		2-1-0	1-8-4	0-6-0	1-0-2
1/2		0-10-7	0-4-4		1-1-2	0-1-7	0-7-5
1	1-7-4	1-0-3	0-8-1	1-11-0	0-9-1	0-1-4	0-4-7
1 1/2		1-2-5	1-1-4		0-6-3	0-1-4	0-3-5
2	2-0-5	1-4-3	1-5-0	1-9-6	0-5-0	0-1-4	0-2-4
3	2-4-1	1-8-0	1-9-0	1-8-6	0-4-0	0-1-4	0-2-1
4	2-5-7	1-10-0	1-11-0	1-8-0	0-4-0	0-1-4	0-2-1
5	2-5-2	1-10-2	1-11-2	1-7-3	0-4-0	0-1-4	0-2-1
6	2-2-7	1-9-3	1-10-0	1-7-1	0-4-6	0-2-2	0-2-7
7	1-10-6	1-7-0	1-5-3	1-7-3	0-6-3	0-3-7	0-4-4
T	1-5-0	1-4-0		1-8-0	0-8-4	0-6-0	0-6-5

STATIONS 20" ON CENTER
BOW RADIUS 24½"
TRANSOM RAKE 10"

ALL BOTTOM SECTIONS FROM STA. 3 FORWARD ARE ARCS OF 6" RADIUS.

C.B. 3" AFT #4
191# PER IN. IMM.
.62 P.C.
3.44/1 W.L. L/B
23° BOW HALF ANGLE

QUARTER BOTTOM BUTTOCK
BASE
622#

QUARTER BOTTOM BUTTOCK
CHINE
SHEER
WL
BASE

SEAT HINGES UP FOR ACCESS TO FOREPEAK
ROWING JUMP SEAT
SEAT HINGES UP
3-5 H.P.
STEP

14'6" × 13'4" × 5'0" × 0'10"

WHEELBARROW GARVEY
FOR DR. & MRS. ROBT. DIAMOND
THOMAS FIRTH JONES, DESIGNER
TUCKAHOE, N.J.
DESIGN #29
MAY, 1988

arc bottom allows a deeper, more comfortable boat. The bilge water, if any, will not slosh from side to side. With so much power on hand, and 5 knots the desired top speed, the lower resistance of the arc bottom was not an important consideration.

Bob originally inclined to a plywood boat, which could have been somewhat cheaper. Good marine plywood can make a strong, light, and durable boat. But the framing of a plywood boat collects water and dirt when the boat is stored right-side up on the beach. And the curves of the garvey's bow, wheel, and wheel well are not easily achieved in plywood.

Foam Sandwich

The Diamond's foam-sandwich garvey was not an easy boat to loft or frame, and this shape is not recommended for inexperienced amateur builders. But the building technique is exceedingly simple. Foam sandwich is simpler than glass over ply, and certainly it makes a lighter and more durable boat. In addition to the problem of glass trapping moisture in plywood, moisture can make plywood swell enough to crack glass, unless the glass is very thick. Polyester resin does not adhere well enough to

Figure 2–3. Mold for wheelbarrow garvey. Foam-sandwich wheel well already in place.

wood, and epoxy resin is incompatible with some types of glass, because epoxy does not dissolve the binders in the fabric.

To build a foam-sandwich boat, frames are set up on a strongback. They are not part of the finished product, and can be made of junk, but should be sturdy. Stringers are screwed to the frames, and they should be pretty good, clear lumber, because they must lie fair, even if they're not permanent. Sheets of foam are laid over the stringers and fastened from inside, with sheetmetal screws or stitching. The foam is faired, puttied and glassed on the outside. The fastenings are removed, and the hull is lifted off the mold and turned over. The inside is faired, puttied, and glassed. Paint or gelcoat is put on with brush, roller, or sprayer. The hull is done.

Of the many different foam sheets now on the market, only PVC is worth considering. Some other kinds of foam, such as polystyrene (Styrofoam), dissolve in polyester resin. Others, such as polyurethane, are too weak and crumbly for boatbuilding, though they are cheaper than PVC and don't dissolve in resin, so they might be fine for iceboxes.

The chief problem with PVC foam, which has limited its use in both amateur and professional boatbuilding, is its cost. But foam can be pieced together, so there is less waste than with wood. Broken or short-cut pieces are never discarded. And as always, hull materials are the worst possible place to save money when building a boat.

Airex is the strongest foam on the market, in my opinion, and the nicest to work with. It holds screws almost as well as cedar, and bends any amount without breaking. It is also the most expensive, and will distort at temperatures over 140 degrees F, taking the glass with it. In St. Martins we once saw an English-built trimaran with an Airex deck that had yielded to the tropical sun, and curved like the waves of the sea. In the Diamonds' boat, I used Airex for the bow sections and the wheel rim and well. These parts are protected from the sun by the foredeck.

Other PVC foams such as Termanto, Klegecell, and Corecell are manufactured by a different process. They cost a third less and are almost as strong as Airex in everything but compression. Resin adheres to them just as well as to Airex, and certainly they are strong enough for all but the heaviest loads. They are relatively brittle, and do not hold fastenings well. They don't heat-soften until 180 degrees, so they are first choice for the deck of any boat, or for the hulls of open boats that are left on the beach.

Further savings in PVC foam can be made by using a low-density foam, but I don't know how you'd bend this stuff, or hold it in place until glassed. End-grain balsa is cheaper yet, and is said to be strong and durable. It comes as cut blocks glued to scrim, so it won't span even narrow gaps between stringers. Mass-production builders often use it, vacuum bagging it into the mold between layers of glass. To use it in a one-off boat, the mold would have to be much more elaborate.

If the bend is too sharp for the foam you're using (and even Airex won't bend into a perfectly fair, tight curve), it will do anything after a few minutes in an oven at 225 degrees. Such an oven can be made from plywood and a hair dryer, but keep the foam away from the direct blast of hot air. In this boat, the only curves that needed heat were the wheel rim and well, and Carol allowed me to use the kitchen range.

Fiberglass

Glass is available in rapidly proliferating variety. For a small displacement motorboat or an unballasted sailboat, the glass skins on either side of the foam can hardly be too light, and the structure will be plenty strong. But for resistance to abrasion and localized impacts, a thicker glass is helpful, and needn't add too much weight. The four forms in which glass fiber is commonly available are chopped-strand mat, cloth, woven roving, and sewn or knitted fabric.

Mat *must* be used next to the foam, or a good bond will not be achieved. But mat is not very strong, and a layer of woven or sewn fabric

Figure 2–4. Glass types, clockwise from upper left: mat, cloth, sewn fabric, woven roving.

should go on top of it. Woven roving and sewn fabric are available with a mat backing, known by various tradenames such as Fabmat, so two layers can be put on at once, and that is what I recommend. A single layer of light Fabmat, doubled at chines and bow on the outside, was what went into the Diamonds' garvey, and it would be plenty for a boat twice the size and speed.

Mat is described in ounces per square foot; other fabrics in ounces per square yard. Thus 1708 sewn-fabric Fabmat, which is what the garvey got, is 17 ounces of unwoven rovings per square yard, and 0.8 ounces of mat per square foot. The unwoven rovings soak up about their own weight in resin, and the mat about three times its weight.

Woven rovings and cloth soak up about $1\frac{1}{2}$ times their weight. In addition, the fact that they are kinked by the weaving makes them less strong than unwoven rovings, though still far stronger than chopped-strand mat. Sewn fabric—biaxial roving—is the newest thing in glass. Bundles of rovings are simply laid out at 90 degrees to each other and sewn together with polyester thread. The rovings remain straight, and therefore strong. They are also tighter packed, and therefore take less resin to wet out, making the resulting laminate stronger for its weight or thickness. Sewn fabric costs about half again more than woven roving, but some of that is recouped in the resin saved, and it drapes and wets out much easier than woven roving. Add the working advantages to the superior finished product, and I wouldn't use anything else on foam.

I like biaxial sewn fabric, in which the two layers of fiber bundles are at 45 degrees to the edge of the roll. Sewn fabric can also be bought with one layer oriented parallel to the edge, and the other layer across it. Often the layers are different weights, to make the fabric stronger in one direction. Unidirectional glass is also available. Then, for another step up in technology and price, S-glass can be used instead of E-glass (everything we've discussed so far is known as E-glass). With S-glass, the individual glass fibers are thinner, so are closer packed, and the resultant laminate will have a higher glass-to-resin ratio. More money can also be spent on more sophisticated resins, which are stronger or more flexible or more waterproof than general-purpose ortho resin. All of these materials are needed in the building of some boats, but not in a 15-foot, 5-knot beach boat.

The fun thing about foam-sandwich building is how much you can get done in a day, or even an hour. When the foam is in place and faired (all corners should be rounded, so the glass can curve around them), you roll out the Fabmat and tailor it. Then, when a double check shows that you have at hand every tool and supply you might possibly need, you put on the rubber gloves and fill all the cracks with a putty you've made of resin thickened with colloidal silica and microballoons or Q-cells. This putty need not be catalyzed, because the resin you're putting on next will set it off. Once you catalyze the first batch of resin, it's a rush, and it's disgusting too, with plenty of resin and odd bits of glass on your shoes and clothes and glasses and hair. But an hour later, it's time to wash the tools in acetone (another disgusting substance), take off the gloves, and have a smoke. Wow! The whole bottom is done, and it isn't even lunch time! Maybe we'll launch her this afternoon.

The finished boat doesn't come together quite that fast. After glassing, the weave must be filled with either a very thin mat called surfacing tissue or with a thin fairing putty. I prefer the latter, and put it on with a trowel. A little wax goes into the resin now, to make it go off hard enough to sand without gumming. On the inside of a foam-sandwich hull, I just sand the rough spots of the layup and roll on gelcoat. The pattern of the fabric remains visible, and that's more attractive than the porcelain finish of molded boats.

Wooden parts are attached to foam sandwich by sanding the gelcoat back until the glass is visible, and gluing them with epoxy. Sheetmetal screws can hold them in place while the glue cures. Areas of severe strain, such as shaft strut mountings, should be solid glass, because the foam won't take the compression of the bolts. The outside of the foam is masked, perhaps with duct tape, and when the hull is turned, the foam is gouged away in that spot and the inner glass skin is laid so that it meets the outer. Extra layers of reinforcing glass may also be necessary, but they go on wonderfully easily. Like a metal boat, foam sandwich allows you to add strength at any point, made out of scrap, without fastenings or even much calculation.

The deck of the garvey was made on a table, and then installed on the boat. Foam was laid out flat, and one side was glassed. The foam and single skin were then bent to the desired curve, and the other skin put on. It was made slightly oversize, scribed and cut to fit the hull, and glass-taped in place. The outboard mounts on plywood pads to spread the compression load of its screws over a wide area of skin. On more powerful boats, plywood rather than foam is the inevitable transom core. Glassed over, it is always the first part of the boat to give trouble.

Complete except for motor and Bimini top, the Diamonds' garvey has 105 pounds on the wheel, and 44 pounds on the handles. Seating is amidships, because the crew contributes most of the weight. The boat seats four pretty comfortably, though it's drier with two aboard. The 3-horsepower Yamaha gives an easy 5 knots in either case. The helmsman steers with a tiller extension, and the boat turns pretty quickly, as it tends to pivot around the wheel like a centerboard dinghy. At 5 knots it's hard to get in trouble. The wheel, I'm glad to say, is not noisy under way, and

Figure 2–5. The wheelbarrow garvey, with only 44 pounds on the handles, is readily trundled down the beach.

doesn't need a brake. No doubt it turns some. The axle, a PVC pipe, turns in another section of pipe with about 1/8-inch clearance to let the sand wash out. There is a jump seat so that Bob can row the boat if he feels like it, or balance it if he takes it out alone.

The Diamonds like their boat, and they like the way beach strollers

Figure 2–6. Wheel with axle made from PVC pipe.

look twice when they see it being wheeled down to the water. Bob caught a fish the first time he tried. He is becoming a student of garveys, and stops to look at them whenever he can. Sometimes he admires the shape or the paint or the 100-horse engine, but he always turns away shaking his head. "Nice boat," he will say, "but it doesn't have a wheel."

Garvey Shapes

An enthusiast of traditional garveys would not find much to applaud in the Diamonds' boat. He would find fault with the materials, the wheel and handles, and the arc bottom aft. The V-bottom forward does have historic precedent, but more in Maryland and Virginia than in New Jersey.

In a culture as uniform as America's, it's rare to find an artifact as regional as the garvey. Spot one in Florida, and most likely the owner will be an expatriate from New Jersey or Maryland. And the sailboat original is more recognizable in the present powerboat than is any other traditional type.

Gasoline motors replaced sails much quicker than they replaced horses. In my own childhood, and even later, horses were still much used commercially by hucksters, milkmen, and farmers, among others. But the only unpowered boat I ever saw working in America was a pound-net rowboat nearly 40 feet long. She came in through the breakers near Mantoloking, New Jersey, early one morning in 1940, and was dragged up the beach by horses. To fishermen, even the finicky motors of the first decade of the century seemed enormously more reliable than sails, and enormously more desirable than oars, so they installed them in their boats at once. Commercial sail may have a future internationally, for long voyages in areas of reliable wind. Inshore it has none, except perhaps as an auxiliary.

Chapelle points out that in converting sailing garveys to power, fishermen also went a long way toward converting the hull shape, too. To make room for the motor, the thwarts were removed. The sides then fell in, reducing the rocker. After that, the motor may have been needed as much for the pump as for the propeller, but the shape wasn't too bad for motoring. By the time cheap motors were powerful enough to plane a garvey (and by that time fishermen were rich enough to afford the foolish vanity of planing their workboats), power garveys were being built from scratch, not converted from sail.

In a sailboat or rowboat, a long waterline increases potential top speed, so overhangs tend to be short unless distorted by fashion or a racing rule. In planing powerboats, a long waterline is less useful, so planing garveys often have a longer and more gradual curve to the bow. This allows chine logs to be bent from unsteamed green oak. Carried to extremes, it even allows the bow to be planked with a single layer of plywood. One does see such garveys, but like the flat-bowed ones, they do not change hands for good prices.

At the boatyard in Port Republic, New Jersey, there are still several working garveys 35 to 40 feet long. Typically powered by 100-horsepower automobile engines, they are too big and heavy to go much beyond hull speed. Low and narrow in proportion to their length, fully decked, with sweeping sheers and wheelhouses well aft, they are majestic, and I hope they'll last a while longer. Bottom form varies, though all of them have less rocker than they should. The rockerless bottom is so easy to build that it is often seen on the heaviest boats, even steel boats, that have no possibility of planing. Fuel is so cheap in this country that builders don't have to think logically about such things, and they'd rather not. But even here, dragging a big transom at hull speed costs something to the environment as well as the pocket, and it makes the boat less seaworthy. It's time builders wised up.

Forward, the Port Republic garveys vary from square to V-shaped. In the last days of working sail, a good many garveys were built around Chincoteague, Virginia, with a slight "V" from bow to stern. My own guess is that the extra trouble and materials this cost would have been better spent on a flat-bottomed, longer boat. But as powerboats go faster than sailboats, they hit the waves harder. Going to windward, a power garvey doesn't heel like a sailing one, so the bow hits the waves dead on. In big water, many skippers prefer some "V" in the bow. In New Jersey, these boats are known as "chicken-breasted" garveys.

The chicken breast is sometimes achieved with a bow still square in plan, by bending the keelson more than the chine logs. Sometimes the keelson and chine logs are bent the same, and the V-bottom is achieved with a bow that is V-shaped in plan—like the Diamonds' boat. In one case in Port Republic, an immense deadwood has been added to a flat bow, making it look V-shaped from some angles, but serving no other purpose. To garveymen, the chicken breast suggests a rugged, offshore vessel. Many of them think of themselves as rugged offshore types, so they will often paint a chicken breast onto the bow of a flat-bottomed boat. I do it sometimes myself.

The potential of a garvey as a powerboat has been well exploited. Its potential as a sailboat has been much neglected. If a sailboat is to be double-ended, it had better be square on both ends. Sailboats that are pointed on both ends lack the bearing to carry sail, and must compensate with more ballast or more beam amidships, either of which slows the boat more than square ends. Square bows are noisy through the water, and some sensitive souls may have problems with their looks, but they aren't slow, as lakes scows well demonstrate; and in sheltered water, the scow may be the best hull form of all.

Beach Garvey

The Beach Garvey was designed long ago for a competition that it didn't win. The judges were savvier than I was. But the drawings caught the eye of my cousin Chris Mayer, and he decided to build one. In it, he made a brave voyage from Alexandria, Virginia, to the mouth of the Potomac and back—a good 200 miles. In two long weeks, he broke his mast and built a new one, and met all kinds of people. He enjoyed the boat, and we both learned from it.

The competition called for a one-person camper-cruiser, powered by oars and sail, with a hull light enough to be carried up the beach by the skipper alone. There may have been other fatuous complications, like spars that stowed within the hull. I took my departure from the Garvey

Figure 2–7. Cousin Chris in his beach garvey.

Box, smallest of the garveys in Chapelle's *American Small Sailing Craft.*
The math was simple enough:

complete boat	150 pounds
crew	175
gear and stores	150
total	475 pounds

I did not alter Chapelle's length-to-beam ratio, so critical for low resistance, especially in chined boats. I increased the rocker to gain the desired displacement, narrowed the bow and arced the stern for a smoother entry and exit and a lower PC, which in garveys tends to be too high for the speed usually achieved under sail. I raised the freeboard as much as I dared, and decked nearly the whole boat.

Both the decking and the small, low rig were intended to protect the precious cargo—bedding, food, books—from any possibility of capsize. At the time I thought the standing lug rig had less heeling moment than other rigs, but I've changed my mind since. At any given wind speed, a square foot of canvas is pushed by the same force, no matter what its arrangement. In a more efficient sail like a spritsail, a larger part of that push is used for lift, driving the boat forward. But in a lugsail, practically all of it is drag, heeling the boat over. The daggerboard is set far off center to give a lying-down space, and the rest of the accommodation is left to the imagination or hardiness of the crew.

In those days, I was keen on building boats without a strongback, and the beach garvey was supposed to go together that way. Longitudinals were on the outside of the topsides, and were glued to them before they were bent around three temporarily braced bulkheads and the transom. Chris found this arrangement impossibly wiggly, and decided to finish the boat on a strongback. To keep the weight of the bare hull down to 100 pounds, which seemed enough to lift, $\frac{1}{2}$-inch × $\frac{3}{4}$-inch strip planking was specified for the bottom. The bulk of a boat like this makes it difficult to lift, whatever the weight, and as Chris was planning to trailer, I told him $\frac{3}{4}$-inch-square strips would do no harm. To save the price of the wood, he went ahead with the thinner bottom, and had much trouble with it later.

The sail came from my old friend Gilbert Webster, who made me many good sails in those years. Gil began as a skipjack sailmaker in Deal Island, Maryland, but moved up to Philadelphia to reap the economic fruits of World War II. Thirty years later, he was still making tents and thousands of tent ropes for the Army. "Always in demand," he said. "When they strike the tents, they cut 'em off." When I knew him, Gil didn't chew, but he must have once, because he always talked like a man with a mouth full of tobacco. I once asked him if he liked working for the gov-

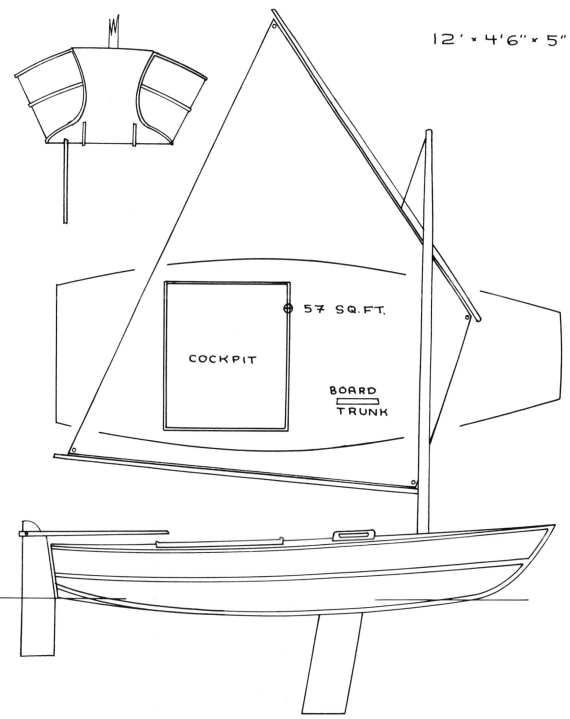

12' × 4'6" × 5"

57 SQ. FT.

COCKPIT

BOARD
TRUNK

Beach Garvey.

53

ernment. Through the lock of white hair that fell down over his forehead, his nearly blind eyes nearly focused on me. "I like to get a gummint check!" said Gil.

From the sweepings of his loft, he made Chris a very cheap sail, as gorgeous as Joseph's coat. The panels varied as much in weight as in color, and perhaps there were nylon as well as Dacron cloths. In a sail that took a shape, it would have mattered a good deal. In a standing lug it probably didn't, and Gil thought it was all pretty funny.

Chris kept the Beach Garvey several seasons. He made more than one voyage in it, and did considerable daysailing as well. It was always dry, and was reasonably well balanced, tacking and jibing reliably. But compared to a boat designed for daysailing, the Beach Garvey was pathetically slow and unweatherly. He also found that, when he sailed it light, it was both slower and cranker than when loaded with two crew, or with one crew and a week's cruising gear. Lightening the boat shrank the wetted surface, but it also shrank the waterline length much quicker than the beam, and it was troublesome to keep the boat on course.

Chris varnished the inside of the white pine bottom, and sanded and revarnished it every spring. I doubt there was 1/4 inch left in the end, and much of the water that the freeboard and decking were designed to keep out found its way in through the bottom. When he sold it, we were all relieved.

The competition winners were all longer, narrower boats than the Beach Garvey, and I now agree with that solution, if you *must* have a sailing and rowing cruiser that you can backpack. However, a couple of inflatable rollers would expand the possibilities of such a boat all out of proportion to the space they took up inside her.

Platt Montfort has lately been working with an improvement to the old canvas-over-stringers method of boatbuilding, using heat-shrunk Dacron, with diagonal Kevlar rovings to stiffen the structure. From the weights he reports for smaller boats, it would seem that a 100-pounder could be 30 feet long, and have a hot tub made in the same manner. If I had another crack at this competition, I'd probably design for Montfort's method, in a boat about 20 feet by 4 feet with a two-masted rig to free up the center of the hull for living space. But really, canoes and kayaks—not rowboats—are the best answer to a cruising boat that can be carried.

12-Foot Sailing Garvey

This boat took advantage of the lessons learned in the Beach Garvey. Length and beam are about the same, but freeboard is less, and so is rocker, giving a displacement of 387 pounds. Carvel built, the hull weighs 153 pounds, and the whole boat with floorboards and rig may weigh 195. Soakage—and carvel boats want to be kept in the water—is probably another 50 pounds. There is little danger of this one sailing above her marks.

The biggest change is the enormous rig: 90 square feet on a 10-foot waterline. Even at the best of times, sailboats are so slow that a slow one isn't fun at all, but this one is plenty of fun to sail. Her second owner did capsize her once, sailing her at the shore in the strong afternoon breeze. He grounded the board, and over she went. In four seasons of sailing her on the Tuckahoe, it never happened to me. Her weight and chines made her less skittish than most daysailers, and her rig more than made up for the weight. Cousin Chris is now her third owner. He keeps her tied behind his houseboat in Alexandria, and daysails her on the Potomac.

She draws much surprise and admiration. ("A sailing garvey? Never heard of one!") Roger Allen, who runs the boatbuilding workshop at the Philadelphia Maritime Museum, came down to see a couple of catboats I'd built (see Chapter 5), but was much more taken with the garvey. After looking through Chapelle, he decided on the 17-footer on page 63, and had the workshop turn one out. They made their usual furniture-quality job of it. They changed no dimension of the hull, but split the single-sail rig by adding another mast, and hid an outboard in a well. Roger claims the boat sails well, but I know my 12-footer would sail circles around it.

Most people who have bought plans and built this garvey have trailered their boats, and have used a cross-strip bottom to keep them tight. For myself, I wanted a carvel boat. Even if kept in the water, carvel boats are seldom perfectly tight, and the garvey was no exception. The bottom swelled up fine, but the topsides dried and opened, and when the boat heeled, water trickled in. Second growth wood, less dimensionally stable than virgin forest wood, aggravates this problem, and even modern stickum doesn't entirely close the gap. Carvel boats need some maintenance, but most skippers like to work on their boats, which is why you see them industriously polishing their fiberglass hulls, which only shortens the gelcoat's life, or hosing the saltwater off their wood decks, which only accelerates rot. But carvel boats are wonderful fun to build, and they can stand bashing around better than most wood or fiberglass boats, because the skin is so thick you hardly ever bash right through it, and if you do it's not much trouble to stick in another plank. Some liveries still use carvel boats because they're easiest for the summer help to repair.

Carvel planking should always be edge-grain, which is to say either quarter-sawn or taken from plain-sawn flitches that include the center of the tree. Wood shrinks and swells more along the annular rings than across them. With cedar or other knotty woods, edge-grain planks can have knots that run the whole width of them, perhaps only peeking out

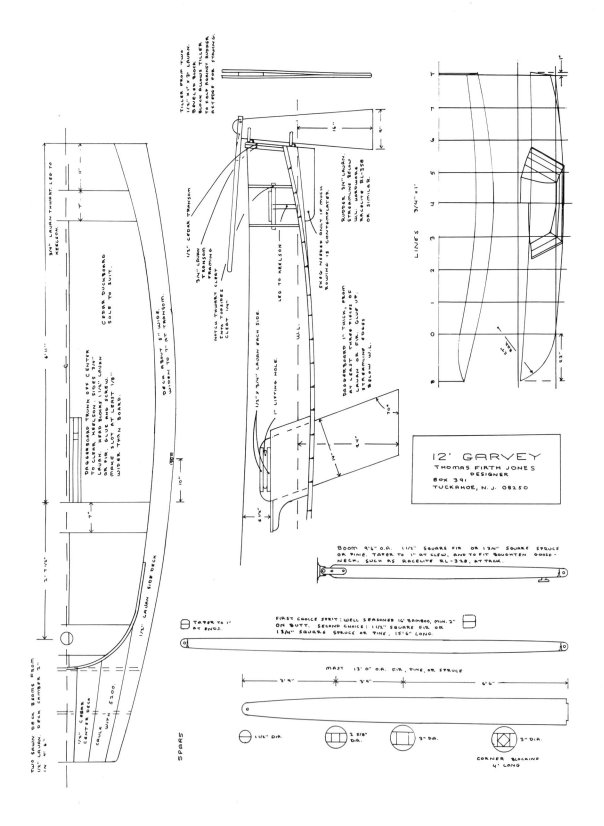

TILLER FROM TWO
1¼" X 1" X 3' LAUAN.
BEVELED BLOCK.
BLOCK ALLOWS TILLER
TO FOLD AGAINST RUDDER
AT EDGE FOR STOWING.

¾" LAUAN THWART. LEG TO
KEELSON.

1"
7"
T

½" CEDAR TRANSOM

¾" LAUAN
TRANSOM
FRAMING

NOTCH THWART
INTO TOPSIDES
CLEAT ¼"

LEG TO KEELSON

SKEG NEEDED ONLY IF MUCH
ROWING IS CONTEMPLATED.

RUDDER, ¾" LAUAN.
STREAMLINE BELOW
W.L. HARDWARE
RACELITE RL-358
OR SIMILAR.

16"
9"

DAGGERBOARD TRUNK OFF CENTER
TO CLEAR KEELSON. SIDES ¾"
LAUAN. HEAD BLOCKS 1½" LAUAN
OR FIR. GLUE AND SCREW.
MAKE SLOT AT LEAST ⅛"
WIDER THAN BOARD.

CEDAR DUCKBOARD
SOLE TO SUIT.

DECK ABOUT 5" WIDE.
WIDEN TO 1' AT TRANSOM.

1½" X ¾" LAUAN EACH SIDE.

1" LIFTING HOLE.

DAGGERBOARD 1" THICK, FROM
AT LEAST THREE PIECES OF
LAUAN OR FIR. GLUE UP.
STREAMLINE EDGES
BELOW W.L.

70°

9'4"

14"

6½"

LINES ¾" = 1'

0 1 2 3 4 5 6 7 T

22"

12' GARVEY
THOMAS FIRTH JONES
DESIGNER
BOX 391
TUCKAHOE, N.J. 08250

BOOM 9'6" O.A. 1½" SQUARE FIR OR 1¾" SQUARE SPRUCE
OR PINE. TAPER TO 1" AT CLEW, AND TO FIT BOUGHTEN GOOSE-
NECK, SUCH AS RACELITE RL-328, AT TACK.

TAPER TO 1"
AT ENDS.

FIRST CHOICE SPRIT: WELL SEASONED 16' BAMBOO, MIN. 2"
ON BUTT. SECOND CHOICE: 1½" SQUARE FIR OR
1¾" SQUARE SPRUCE OR PINE, 15'6" LONG.

MAST 13' 0" O.A. FIR, PINE, OR SPRUCE

3'9" 3'9" 6'6"

1½" DIA. 2 5/8"
DIA. 3" DIA. 3" DIA.

CORNER BLOCKING
4' LONG

TWO SAWN DECK BEAMS FROM
½" LAUAN. DECK CAMBER 2"
IN 4'. 6"

2' 7 7/8"
6' 11"

7"
10"

½" CEDAR
CENTER DECK
CAULK
WITH 5200

½" LAUAN SIDE DECK

SPARS

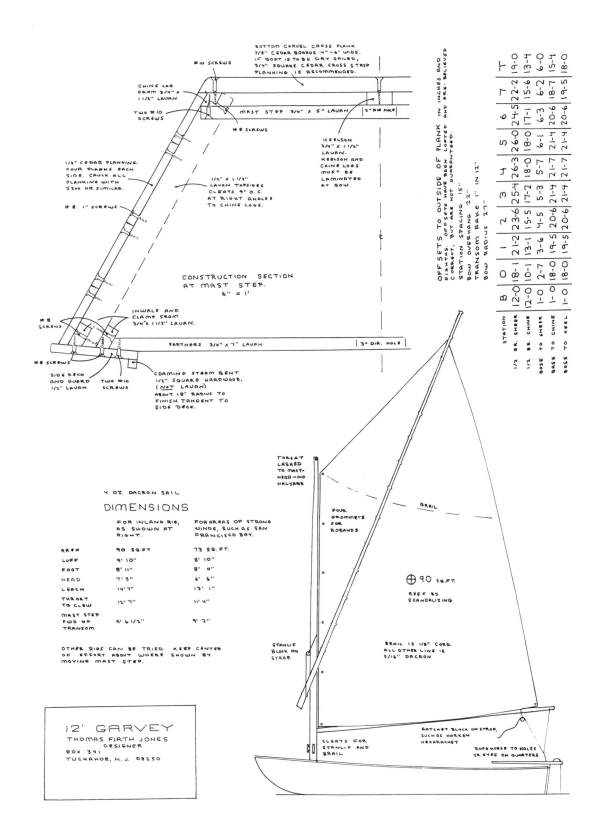

Figure 2–8. 12-foot garvey under sail in the Tuckahoe.

at the two edges. So carvel planking must be chosen with greater care than clinker planking, which can shrink and swell some if it wants to without opening up the boat.

The sailing garvey was built upside down on a strongback (Figure 2–9). Four molds and the transom were set up. The bow framing was laminated from mahogany veneers 3 inches wide, and ripped to give two chine logs and a keelson. Topsides cleats were notched but not beveled, and were screwed on at right angles to chine log and sheer clamp. Topsides planking, ½-inch cedar, could have been spiled by someone wanting the practice, but parallel-sided planks fell nearly parallel to chine and sheer. Thwarts and daggerboard trunk went in before the bottom went on, with the trunk offset just enough to clear the keelson.

At the bow, the bottom planks had to be narrow and backed out to take the spoon curve. At the stern, the ⅞-inch planks needed firm twisting

Figure 2–9. Sailing garvey in frame. Molds still in place.

and meaty screws to conform to the V-bottom. If flat-bottomed, this boat would certainly have planed in ideal conditions, but she wouldn't have tacked as quickly or sailed as fast in most conditions. In a narrow river only three miles long, planing is not desirable because it uses up the river too fast, and there's no chance to relax and smoke. "If you can't smoke while you're doing it, it isn't a sport," says Carol.

Before taking the boat off the strongback, I drew the waterline and kerfed it in with a panel saw. A special hook-bladed pocket knife called a rase knife was once available to boatbuilders who did a lot of waterline cutting.

Side decks were mahogany, $1/2$ inch thick, and overhung the topsides by $1/2$ inch to make a guard. The line to be cut was marked on top of the deck with an underscriber, a tool easily made and often used by Formica workers (Figure 2–10). Forward, the center deck planks were $1/2$-inch cedar. The coaming was steam-bent, which was probably a silly vanity in a boat so boxy.

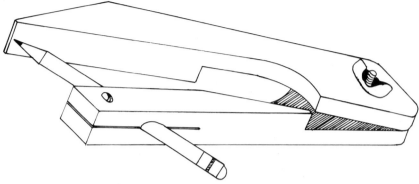

Figure 2–10. Underscriber.

Sprit Rig

To windward, the sprit rig is probably only 80 percent as efficient as a Bermudan rig would be, but this one does have the extra 20 percent of area to compensate. Downwind, all rigs seem about equally efficient, and sail area is never large enough. I've found a good spritsail far superior to any other low rig, including the low-aspect-ratio Bermudan. To give good lift, a jibless spritsail needs an aspect ratio of about 1.5. Contrary to what Chapelle and others have written, I have not found that the head of the sail needs high peaking: 45 degrees is probably best, and peaking it up more lengthens the sprit, or shrinks sail area. It also brings the sprit too close to being parallel with the head, which makes sail adjustment too exacting.

Like any other sail, a spritsail needs a boom or else it will have too much camber too far aft. Downwind in strong conditions, boomless sails pump, augmenting rhythmic roll. Spritsails also work better with battens, though this garvey never had them. (Our sail was the cut-down mainsail of a catamaran. See Chapter 7.) Battens allow a roach, which is a desirable but not overwhelming addition to sail area. More importantly, they help control the shape of the sail. Much of the ill repute of spritsails and other low rigs came about because their skippers belonged to the "hang up the rag" school of sailing. All sails need to be well designed and cut, and much studied and fussed with.

Finely tuned spritsails can be seen in quantity at the better Optimist Pram regattas. It's no easy trick for them, because by class rules the Optimist sprit is too short. Sprits should be at least twice as long as the sail head. The sprit heel should be controlled by a stanliff, not just hung in a snotter, and in a spritsail any bigger than the garvey's, the stanliff should have a multipart purchase.

Compared to other four-sided sails, a spritsail is easy to adjust. It can be very lightly sparred. It twists very little, and does not chafe. Up to the point where the sprit becomes too big to handle, it is a wonderful sailing rig.

In the garvey, we handled the sprit infrequently, and in any case it was light, being a 16-foot stick of Jersey-grown bamboo, 2 inches at the butt and weighing less than 2 pounds. Seasoned for five years before use, it was plenty strong, though it did bend some in high winds. It lasted seven or eight years, but recently Chris found it splitting (probably from sunlight) and replaced it with a wooden one.

The luff was tied to the mast with individual robands at each grommet, and this is the best method. Hoops are an affectation in boats under 40 feet: They are expensive in time and money, noisy in stays, and destructive of the mast finish. A continuous mast lacing, no matter how tricky, guarantees an ill-setting sail. The largest gaff- or spritsail I ever handled regularly was 150 square feet, but from that one down to the smallest, the throat roband always worked best without parrel beads. The garvey had no halyard; the throat was simply tied to the masthead. To allow fine adjustment, one line was needed to hold the sail up, and another to hold it to the mast. The same is true of outhauls, especially on loose-footed sails.

The garvey was kept in the water, and the rig was kept in the shed. The boom was in the boat, and to get under way, I carried the rig down; stepped the mast; hooked up the gooseneck; and released the brail line, which had gathered sprit and sail to mast. The sail dropped, and I cleated the outhaul to the boom and sailed away. It was so quick and easy that, when the tide was right, I sometimes sailed most days of the week.

Sailing was quick and easy, too. Starting out, the weather would sometimes require a slight adjustment of outhaul or stanliff, but after that, only the sheet and tiller required attention. With such a big rig, the sheet was never cleated, but with two-part purchase and ratchet block it wasn't too tiring to hold. The boat would tack in 90 degrees, but made its best speed to weather tacking in 100 degrees. This is true of most small daysailers, and a really small one like the El Toro (see Chapter 3) is best tacked in 110 degrees and not pinched at all. An exception is the Laser, with its very efficient sail sleeved over the mast; but the Laser is an exceptional boat in every way, and I'd recommend it to anyone who didn't mind its plastic feel, and didn't mind sailing *on* a boat, rather than *in* one. For myself, sailing in cold as well as warm weather, being inside a boat is important.

The garvey is no slouch downwind, but upwind is her best point of sailing. The importance of upwind performance cannot be overemphasized. The shorter the time spent sailing, and the smaller the water one sails in,

Figure 2–11. Sailing garvey with brailed rig.

the more this is true. An unweatherly boat simply never gets to weather, so never gets to sail any other course. Hobie Cats and Windsurfers, which are better reaching than they are upwind, are usually seen in wide bays, where they can reach across and reach back. Their skippers claim to find it interesting.

The garvey is best to weather because the wetted surface is so much reduced. Upright, with board down, wetted surface is 34 square feet, which isn't bad for 90 square feet of sail. Heeled to the gunwale—and the deck means there's still considerable stability beyond that angle—the underwater shape is an almost equal-sided "V", and the wetted surface is 30 square feet. Waterline beam shrinks from 40 to 30 inches, and waterline length is probably 11 feet. She goes, and she goes quietly, too. The square bow only mumbles when the boat is upright.

Heeled to that point (about 30 degrees) the skipper is well braced and comfortable, with the lee topsides a convenient distance for a foot brace. In lighter winds when the boat heels less, there is still a performance advantage in getting one chine out of the water. The neutral helm at any degree of heel is a revelation. As most boats heel, the underwater shape becomes cockeyed: It's asymmetrical fore and aft as well as athwartships, and the boat is desperate to head up. For some reason, this is even more true in a boat with two pointed ends. But the square bow of the garvey keeps this from happening.

The one problem with a spritsail is that, in tacking, the luff rotates around the mast. The sprit rotates less, however, so peak-to-tack tension on the sail is different on the two tacks. John Leather, in his book on spritsails, says this is supposed to be solved by the mast rotating in partners and step; but in practice I've seldom seen it work, and then only on the tiniest sails, like kayak sails, and only after some delay. It's more likely to work jibing than tacking. Friction and sheet tension both hinder it. A better solution might be found by inducing the stanliff to rotate with the sail, but I haven't figured out how this could be done simply.

The garvey is a disappointing rowboat. Oarlocks are $4\frac{1}{2}$ feet apart, and we used 8-foot oars. The rowing seat is low to give 8 inches height between seat and locks, and well forward to give a comfortable space for two sailors to move around in the bottom. A vigorous pull on the oars brings enough water up the daggerboard trunk to give the oarsman a wet crotch, so a trunk plug is needed if serious rowing is planned. Her skeg, which is superfluous under sail, has to be immersed for her to track at all under oars. To immerse it, rudder and daggerboard are minimal ballast in the stern sheets, and a second person is really better.

Strictly speaking, there is no such thing as a good "combination" boat. Hull shape requirements for sailing, rowing, and motoring are so different that there is no reconciling them, and they can only be compromised. The square ends of the garvey, which help move her center of buoyancy to leeward as soon as she heels, are too blunt to push through the water with oars. She is a sailboat that can be rowed, not a sailing-rowing boat.

Like most small daysailers, she sails best with one crew, but is per-

Figure 2–12. Sailing garvey under oars. Tuckahoe Catboat in background.

fectly tolerant of two. Even with three hefty people aboard, doubling her designed displacement, there's still some fun in sailing her. The Bruce number is a handy yardstick for estimating what sailing performance will be in fresh winds, not in ghosting conditions where sail area–to wetted surface ratio is the best yardstick. Bruce number is calculated by dividing the square root of the sail area in feet by the cube root of the displacement. An average cruising boat has a Bruce number of 1.0; an average daysailer somewhat higher. At 387 pounds, the garvey's Bruce number is 1.30. At 774 pounds, it's still 1.03. She still trundles along, without too much bow wave or wake.

I remember best the hundreds of times when I sailed the garvey alone, beating up the Tuckahoe against the nice west wind we often get in the afternoons. The tide is still flooding, and I'm making windway on it. Momentum gives something in hand for the frequent wind shifts, and she feels like a big powerful boat in this small water. The gunwale is just touching the water, and the sprit is near to bending and creasing the sail, but for now everything is holding. The tiller is amidships and neutral. And she's moving like a train.

3. Daysailers

I learned to sail at Tabor Academy's summer camp, in the years just after World War II. Life jackets then were bulky, likely to leak kapok, and not much in fashion, so the camper had first to pass a swimming test. Then he got a couple of hours of instruction a day, in a boat with a counselor and one or two other campers. Once two campers could pass a sailing test—tacking, jibing, bringing the boat up to the dock firmly enough to crack an egg, but not break it—they could take a boat out alone each day, and race it on Sundays.

Marion harbor was not then a vast parking lot of unused and unwanted but everlasting cruising boats. Near the town dock were a few dozen moorings, some occupied by local boats and quite a few by transients. We always sailed among the moorings, to see who had come or gone. The rest of the harbor was unencumbered, and there was plenty of room for a race course.

The boats available to us were a good number of Beetle Cats, and half a dozen Zips. My crew and I preferred the Zips, because the competition was less intense and the main-and-jib rig gave each of us something to do. They were unballasted, chined keelboats about 12 feet long, heavy and logy, incredibly slower than the Beetle Cats. Our strategy for winning races—and it usually worked—was to slip half a dozen sox into our pockets before being taken out to our boat by the launch. Then, at the mooring, while one of us pretended to be busy with the sails and rigging, the other would secretively slip all the sox onto one hand, one on top of the other, and scrub around under the boat as far as his arm could reach, knocking off the barnacles. This gave us the speed edge on the other Zips, though not on the Beetle Cats.

Though I favored them then, the Zips now seem to me in all respects the antithesis of what a daysailer should be—heavy, complicated, stodgy boats. To Tabor's credit, it was phasing them out even in 1947, and didn't waste much time maintaining them. A good daysailer should be light, simple, and above all responsive. The 12-foot garvey is all those things, and this chapter is about four other good daysailers.

The Dobler

For the last 25 years, Joseph C. Dobler has been drawing innovative and often excellent designs, both monohulls and multihulls. He has won at least one important design competition, yet his name is not always recognized, even among students of yacht design. Dobler is a dour man. Though he writes good letters, he seldom writes for magazines. He seems to have little interest in promoting or publicizing his work.

His "16′ × 5′2″ Utility Skiff," as he somberly calls it, was the first daysailer that Carol and I owned together. Compared to a Lightning, which was the last daysailer I'd owned, the Dobler was marvelously simple, easily managed, and seakindly. She could carry seven people, with room for all of them to move around. With a Sunfish rig—we tried a variety of rigs in her, over the years—she could pull steadily away from a Sunfish on any course except dead upwind, where the low freeboard and windage of the Sunfish hull made the difference. In a calm and with 8-foot oars, she rowed very sweetly. Dobler claimed she could handle a small outboard too, but we never tried one. She was the best compromise I've ever seen of the conflicting needs of sail, power, and oar—largely because she was bigger and simpler than most designers would have drawn her.

With two people aboard, the Dobler sat on her marks. Length-to-beam ratio was 4 to 1; wetted surface was 38 square feet; and the hull weighed just over 200 pounds. People were amazed to see such a big boat go so fast with such a small rig. Sailing around Cape May Harbor in the evening, which was the way we most often used her, the Dobler could pull away from almost any cruising boat under 30 feet. A friend with a Laser once set out to catch us, and eventually he did, but it took him a lot longer

Figure 3–1. Carol rows the Dobler in Cape May Harbor.

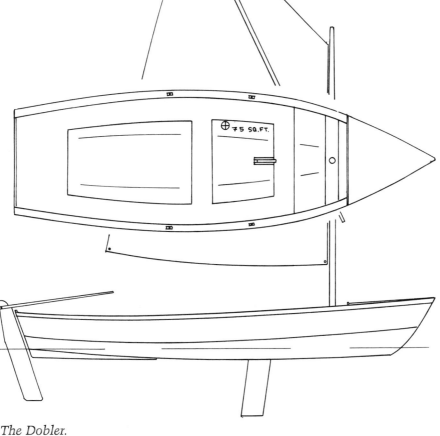

75 SQ.FT.

The Dobler.

than he'd expected. And a local man, the owner of a 37-foot Novi schooner who thought himself a budding Richard Maury, used to grimace when he saw us approaching.

The many seats, at a comfortable height above the bottom, allowed the crew to trim ship for any condition without having to squat on the bottom. Lunch, sweaters, and any amount of gear stowed handily under the stern sheets or the forward deck. The only trouble the boat ever gave was when a contentious drunk at our marina came stomping across the dock. With the idea of punching me in the mouth for some imagined offense, he boarded us by stepping on the gunwale. My mouth went unpunched.

The Dobler was a kind of Swampscott dory in plywood, but probably better than any round-sided Swampscott. This was because she was lighter, and because the chines gave a narrower waterline and more reserve buoyancy above the waterline than rounded topsides could have — though round sides might have looked prettier. Scantlings were light: 3/8-inch bottom and 1/4-inch topsides. Dobler called for a stringer along the middle of each topsides panel, and taped seams at the chines. Mistrusting taped seams in those days, I used chine logs of 3/4-inch mahogany, so the boat probably would have been another 10 pounds lighter if built strictly to plans.

The Dobler plans showed a high-aspect ratio lateen rig, with the lug sleeved over the yard. After I sent Joe Dobler a batch of pictures, confessing that we weren't using his rig, I never heard from him again. Too bad. The different rigs we used had different centers of effort, and we did move the mast step and partners when we changed rigs, although not very scientifically. The mumbo jumbo of finding centers of effort and lateral resistance and selecting a lead among the many and conflicting

---

Plywood

Probably the majority of plywood boats in this country are built of exterior-grade plywood. This thrifty practice is the sole reason that plywood boats have a bad reputation. Exterior ply is not suitable for boats because it has voids, where moisture collects and rot begins; because much less glue is used than in marine panels; and because the veneers are less well-finished before the glue is applied. The plies are stuck to each other occasionally here and there rather than continuously across the sheet. Species of marine ply commonly available in the United States are as follows:

Douglas Fir (Specific gravity, .48). These attractive panels are 60 percent more expensive than A-C exterior plywood, and you'll never spend your money better. They have more plies, fewer voids, fewer patches, and more glue. Varnished, they have a lovely color and pattern. However, they check badly in sunlight, and are not suitable for boats kept outdoors.

Lauan (S.G., .41). This is the species out of which underlayment ply is made, but like Douglas fir marine, lauan marine has fewer imperfections and more glue. It has no more veneers than fir marine, and the face veneers are alarmingly thin, but seem to last okay. It's the cheapest plywood suitable for boats kept outdoors, and we built the Dobler of it. Cost is about 1.2 times that of marine fir.

Okume (S.G., .37). Light weight and flexibility make okume first choice for tortured plywood. However, it is less rot-resistant than other

possibilities is a much overrated pastime, especially in small open boats where a slight change in crew position radically changes the lateral resistance.

The various rigs we tried in the Dobler—two different spritsails, two lateens, and a standing lug—used an unstayed mast and set no jib. I'm convinced the only purpose of a jib in a daysailer under 16 feet is to keep the crew busy. In larger boats, jibs keep individual sails to a manageable size; and in offshore boats, one sail can control the boat while the other is reefed or changed. A perfect contrast to the 16-foot single-sail Dobler is the 11-foot Mirror Dinghy, with its 49-square-foot main, 20-foot jib, and 65-foot spinnaker. The Mirror rig was designed to teach kids the teamwork and sailhandling they would later use on larger boats, and you seldom see adults sailing Mirrors for pleasure.

Handicapping rules have long acknowledged that a single sail gives more drive than two. When development classes are established in which sail area is limited (e.g., the Moth, or the various catamaran sizes), the

---

species, and end grain must be carefully sealed. Like the two species that follow, okume panels are excellently manufactured, with all plies uniform and less than $1/16$ inch in thickness. Cost 1.9 times marine fir.

Khaya (S.G., .44). This is my current favorite (although I recently discovered that its dust is particularly toxic—an argument for a very good respirator), wherever the weight can be borne. A dense, stiff, African hardwood. Cost 2.1 times fir.

Sapele (S.G., .57). A much-touted species, but very little stronger than khaya, and cripplingly heavy for most applications. Bruynzeel, the most famous name in marine ply, makes okume and sapele panels, which they call *regina* for some reason. Cost 2.2 times fir.

Plywood veneers are made by turning a log against a knife. They are flattened, and sheets of glue (available to the manufacturer in various thicknesses) are layered between them. The layers are pressed together, and heat kicks off the glue. All marine ply is made from virgin forest logs, and when these are gone, it's hard to imagine how marine plys will be made.

Today, the panels are available from a number of suppliers who advertise regularly in boating magazines and ship the panels anywhere by truck. When a friend tells you he can't get marine plywood, he's really saying he's too shortsighted to pay for it. He's saying that no matter how much he spends on the replaceable accessories of his boat, he's going to save money on the irreplaceable hull. The longer you are acquainted with his boat, the more you will pity his folly.

Figure 3–2. The Dobler in the garden where she was built, with the specified high-aspect–ratio lateen rig.

earliest examples will often be main-and-jib boats, but the evolution toward a single sail is inexorable. To windward, whatever a jib does to the air flow over the mainsail is done better if the jib's area is added to the main itself. Off the wind, and especially downwind, the superiority of a single sail is even more striking, though a big enough spinnaker can add interest to what is often a dull point of sailing. If a jib luff cannot be kept reasonably taut, the sail provides little help to windward, and the strains that a taut luff and shrouds impose on a hull often call for a stronger, heavier boat. The larger the jib, the more this is true. I doubt that the Dobler could have stood a main-and-jib rig without substantial stiffening and weight gain, and I know it wouldn't have improved performance or the pleasure of sailing her.

Joe Dobler's lateen rig never had a fair trial with us. Our spars were of poor stock and too heavy; the sail was cut down from an old Bermudan main. This is not the way to get fun out of sailing, or to learn about it. Nevertheless, we were dubious about the rig from the start, and what we saw in action didn't encourage us to go first class with it. Then we went to a spritsail, which gave the usual dependable results, though it wasn't quite big enough.

We next tried a Sunfish lateen rig, buying a new sail and building wooden spars. This satisfied us for several years. The Sunfish rig sets up quick if the sail is kept furled around the spars. It should be stowed in a box or bag where the sun can't reach it, however, since its nylon fabric deteriorates in sunlight faster than Dacron. Most capsizes in real Sunfish happen when a gust heels the boat. Sail cannot be eased to right it, because the long boom is already in the water. In the Dobler, the taller mast and greater beam and freeboard kept that from happening.

Eventually, we tired of the Sunfish rig, which can't be much adjusted under way, and tends to be too flat in light air and too full in heavy. We tried a standing lug briefly, as described in the next chapter, and finally

Daysailers

Figure 3–3. The Dobler with her final spritsail.

settled on another spritsail, exactly the size of the Sunfish sail. Even though boomless, it gave the most drive and pleasure of all.

The third owner of our Dobler kept her on the beach in Delaware Bay. He taught his kids to sail in her, and used her as a tender for his larger fishing boat, which was moored offshore. At last, a storm picked up the big, light Dobler, wafted her inland 75 yards, and smashed her on a seawall. Her great volume and light weight, which made her such a delightful and useful boat, were her downfall in the end.

Barrel of Monkeys

The first attempt to build a daysailer for our narrow, winding three miles of the Tuckahoe River turned out to be quite a good and interesting boat—once the bugs were worked out of her. At the time, I was much influenced by Phil Bolger's writing (if Bolger sold used cars, he'd be a billionaire), and particularly his 7-foot 9-inch pram *Fieldmouse*. He pictured this boat in water even smaller than the Tuckahoe, and gave her plumb transoms at both ends, clinker plywood planking, and a stupendous 74 square feet of fully battened Bermudan sail.

Bolger thought of keeping *Fieldmouse* on a cruiser's deck, but we ex-

Figure 3–4. Barrel of Monkeys *in her eleventh season. The sail is older.*

pected to keep *Barrel of Monkeys* in the water, so neither short length nor light weight was quite as important to us. Nine-feet six-inches seemed the right length for the 12-foot building space available, and made good use of 10-foot planks. Three-quarter-inch glued cedar strips promised a reasonably light and strong boat, with no frames to clutter the interior.

A hull with two flat, plumb transoms is the easiest of all to loft. How far to go with it? Lofting is an ancient art, and if it adds pleasure to build-ing a pleasure boat, it could hardly be sinful. But in a hull this size and shape, very few offsets will suffice. I didn't use diagonals in my lofting, but have drawn them on the lines plan. It's hard to see what information they provide that the waterlines and buttocks don't, or how any mold could take a bum curve for the lack of them. Surprisingly few offsets will suffice if the shape is not complicated with reverse curves, and if water-line and buttock spacings are carefully chosen where the sectional curves are greatest. In the end, all you get from the lofting that helps with the set-up are sections and bevels. The other lines only contribute to perfect-ing them.

Lofting and other floor work are hard on the knees. I use a lofting stool, which runs on casters, hangs on the wall when not in use, and has a tray below it for pencils, rulers, etc.

Lines are fascinating for their own sake, but the preoccupation with them is a throwback to the days when people believed that sweet lines were the secret of low resistance. It was once even argued that clinker planking *reduced* resistance, because it directed the water flow over the hull. As famous a designer as Nat Herreshoff carved models instead of

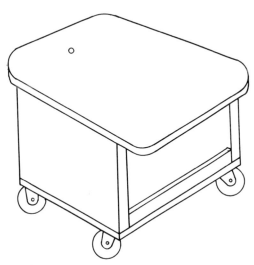

Figure 3–5. Lofting stool.

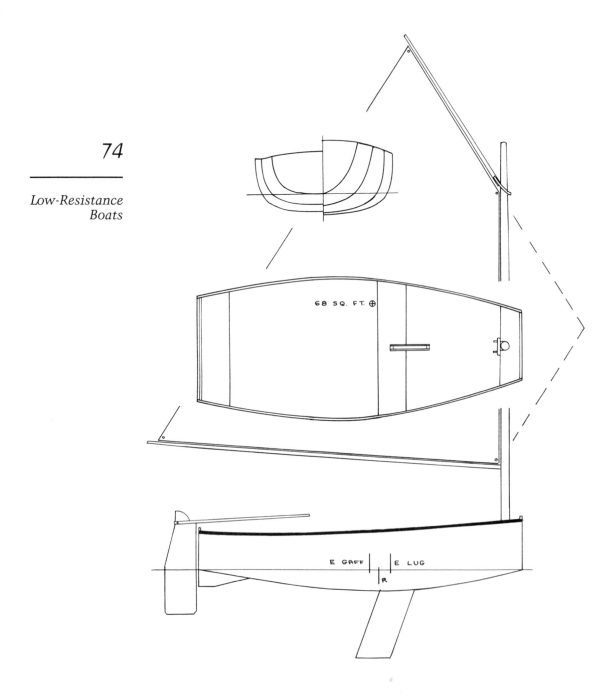

68 SQ. FT.

E GAFF | E LUG
R

Barrel of Monkeys.

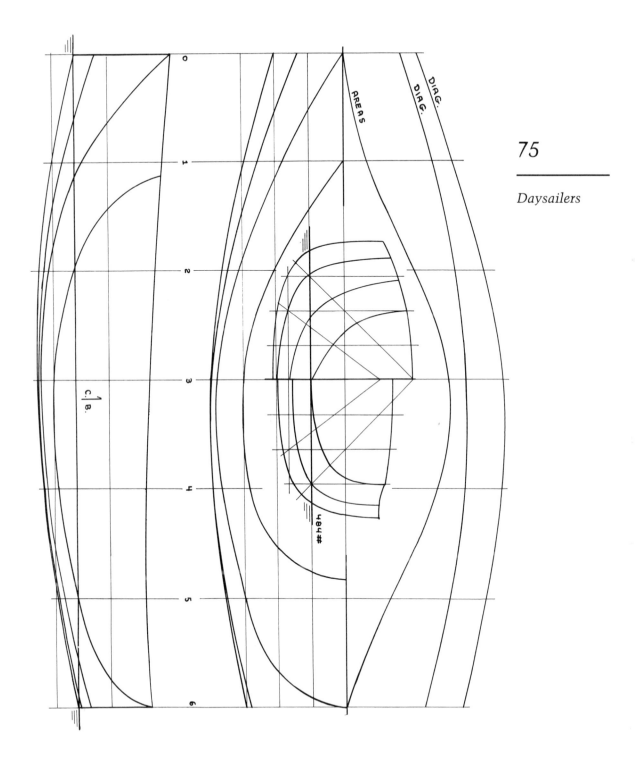

Lines: Barrel of Monkeys.

drawing lines so that he could feel the water flow with his fingers, as well as see it. Herreshoff was a brilliant man, and it is a mistake to think he'd be using the same methods if he were alive today.

Monkeys' lines plan also shows a curve of areas, which is sometimes seen on boat plans from before World War II. Although this curve appears to be linear, it really isn't. In designing a boat, the area at each station must be measured, of course, to calculate displacement. These areas can then be laid out linearly on the stations to some trumped-up scale that has no relation to the scale used in the rest of the drawing, and connected by a curve. The space the curve encloses looks like an area, but is really a volume. A hundred years ago, many designers hoped that the secret of resistance could be found in the curve of areas. Colin Archer spent much intellectual energy modifying the "waveline system." He felt that the curve of areas should be related to the curve of the waves that the boat made in going through the water. This was a good guess, given the information then available, but it happens to be wrong.

The curve of areas does show graphically how full the ends of a hull are compared to the middle. Unlike prismatic coefficient, however, it shows the relative fullness of bow and stern. Nevertheless, today we prefer to describe a hull and the resistance it generates with numbers, not lines. Numerical descriptions are briefer, easier to compare, and less likely to mislead.

Strip Planking

Barrel of Monkeys was built over two transoms and five temporary molds. With planking as stiff as 3/4-inch strips, three molds probably would have sufficed. Though a keel or keelson is sometimes used in strip hulls, I started right in with the cedar, running all the strips full length and tapering them as need be. Shortly before fiberglass became the favorite material for mass-produced boats, strip planking was in vogue, and some yards eliminated the tapering of strips by starting with a bunch of them parallel to the keel. After a certain number had been glued up, a line was struck on them that was a uniform distance from the sheer, making a shape on the bottom planks that looked like a football. This line was cut, and the rest of the hull was planked up.

Some of these mass-produced strip boats have held up well for 40 years now. Others are in trouble, and the trouble is usually the joint between the "football" and the topsides planking. The joint is in effect a scarf, at least for the strips that are inside the football. Minimum acceptable scarf length is 6-to-1 with epoxy glue and 8-to-1 with resorcinol. Usually, the strips nearest the keel wind up much closer to crosscut than those near bow and stern. If a football cut is used, I'd butt-block it heavily in those places, and then I'd be stuck with big hunks of wood on the inside of the hull. For *Monkeys*, the tedium of tapering seemed preferable.

Strip planking is indeed tedious. Carol says: "You're always enthusiastic when you start, but after a few days you come up from the shop muttering about 'little strips.'" At best, it takes me an hour to lay a square foot of strip planking, not counting the time for ripping out the stock. This is much faster than cold molding, but slower than carvel or clinker planking, let alone plywood or fiberglass. What with ripping, tapering, and sanding, as much as a third of the stock you buy is converted to sawdust before the hull is finished, and a considerable part of that goes into your "ears, nose, and other holes," as Nabokov says. And sometimes at the end of the day it's hard to see what progress has been made.

But a strip-planked hull is a lovely thing to see. Twelve years ago, a neighbor bought a 25-foot *Amphibicon* sloop. He built a shed, put the boat in it, and stripped it down to a bare, unpainted hull with the intention of restoring it. We all went to see it, and marveled at how lovely it was. It's still there, in exactly the same condition. And it's still lovely.

Apart from the tedium, I find that strip planking produces the ideal wooden hull. It is light and strong—the 3/4-inch strips in *Monkeys* would do for a 25-footer—and needs no more framing than foam sandwich does. Inside and out, it has a feel that plastic just can't duplicate. The life of a wooden boat is usually thought to be 50 years (25 years for a plywood boat), but I believe that the best strip-plankers will last a lot longer than that without major work. There are no joints or other interstices where dirt and moisture can accumulate, promoting rot. There are absolutely no voids. The hull is monocoque, so it does not work and wear itself out.

A strip-planked hull needs no covering of fiberglass or other goop, although layers of veneer are sometimes put on outside the strips. Builders who do this are usually cold-molders who have grown tired of the technique, rather than strip-plankers looking for extra strength. I can't see the point of adding the veneer, unless a greater hull thickness is required than the strips will bend to, as might be the case with a very short, curved, and heavy hull like a catboat. If the strips won't bend to it, I doubt the hull is a good shape for going through the water. Below the waterline, a thin layer of glass set in epoxy can be a valuable teredo shield. In a cedar hull kept in the water, epoxy glass will retard soakage, and may save its own weight. But the glass certainly isn't needed structurally; and glass over wood, no matter how good the materials and careful the work, usually causes more maintenance than it saves over the years.

Another method of building strip-plankers is to use external molds, and build the boat right-side up. If the shape has deadrise, you start out working a little down-handed, and the farther up the hull you come, the more down-handed you work. But the strips must be held out to the molds (perhaps with temporary screws) and the transom(s) and other bulkheads must be put in after the hull is complete. Perhaps the best method is to use internal molds, hang them from the ceiling, and build the boat right side up, as is sometimes done in carvel construction. *Monkeys* was built upside down, and after the turn of the bilge I got very tired of driving the nails lying on my back. To put an end to it, I cut the molds from the strongback and turned the hull before it was planked.

Turning her over before the topsides could stiffen her, and using less than wonderful materials in the construction (the strips are good cedar, but she is all plastic-resin glued and some of her nails are unbarbed copper), took some of the sheer and rocker out of *Monkeys*. I forget whether the excuse for using such materials was economy or availability, but in building a hull, neither excuse is good enough. Building a boat is a way of *spending* time and money, not of saving them.

As the fastenings have relaxed over the years, a little more sheer and rocker have gone. Designed to displace 484 pounds with both transoms just touching, she may now displace no more than 350 pounds, and she sails much better now with one crew than two, though she has room for at least three. She has also needed maintenance from time to time; seams have leaked, and needed reefing with a hooked hacksaw blade, followed by infusions of epoxy. I don't know how many of her leaks came from construction errors, and how many from the abuse she took from her second owner.

The second owner bought her because he wanted the experience of capsizing a sailboat. In this ambition, I believe he was successful more than once, though he must have worked pretty hard at it. Like most short, round-bottom boats, I always found *Monkeys* would head up before she shipped much water to leeward. The deliberate capsizes may not have hurt her, but he kept her rigged in a very shallow and exposed anchorage. Several times in nor'easters, she capsized unattended and beat herself on the bottom for some hours. When the third owner—an old friend here on the river—brought her home, we found the sheer broken on one side, and had to sister it.

The center thwart created a "hard spot" in the planking. All kinds of hulls can have hard-spot problems, particularly plywood and fiberglass hulls, as well as strip-planked ones. To an extent, the skin is the structure, and it does flex, even if the flexing is hard to see. At the hard spot, it doesn't flex. In *Monkeys*, perhaps from the time when she pounded on the bottom, the thwart eventually opened a serious crack in the topsides, with some distortion of topsides shape as well. In her eleventh season, we spread the load with a sawn frame about a foot long at each end of the thwart, and probably they should have been part of her original construction.

From this catalog of woes, it might seem that *Monkeys* has been nothing but trouble, but she has been used harder than any daysailer I know, and has given service all out of proportion to her trouble. I learned the river in her, and must have sailed her a thousand hours in the three years I owned her. Often I timed my outings provocatively, so that I would be near the railway bridge when Carol and a sailing friend came grinding home on the commuter. It never failed to stop their hearts and stir their ire. With red hull and blue-and-white striped sail and her big, heart-shaped bow transom, she looked impossibly picturesque, but in fact she was a very comfortable and useful boat. She turned quickly, and was perfect for exploring narrow creeks. Her short waterline and rockered bottom held her top speed to 4 knots or so, but her big rig kept her near that speed a good deal of the time. She seemed a real vessel from inside, and she was a responsive sailer, but she didn't use up the small water too fast.

Subsequent owners have used *Monkeys* less, as is often the way with boats, whether they change hands or not. One year, she wasn't launched at all, but every other year she has been in the water at least seven months, and sailed from time to time. Most winters she has been left outdoors, covered with scraps of plastic masquerading as tarpaulins. I've seen a number of fiberglass boats reduced to caricatures of their original shapes and structures from less abuse than *Monkeys* has withstood. Though, in hindsight, she could have been better, she is still a sturdy vessel.

Leeboards

Monkeys started life with a single leeboard. A round-bottom boat should be sailed flat whenever possible, so I figured that most of the time the board would work as well on one tack as the other. Two leeboards would be more trouble, clutter, and weight than a daggerboard and trunk. The board could pivot up for grounding or for running. It was strongly mounted to take pressure from either side. From the waterline down, it was shaped like a leeboard: The section was symmetrical fore-and-aft rather than round in front and sharp in back, as a daggerboard should be. The Dutch were not the only sailors to use the leeboard. Why wouldn't it work?

The leeboard didn't work on either tack. Years later on our *Hummingbird* trimaran (see Chapter 7) I got used to shifting the float-mounted daggerboard when we tacked. I would walk up the windward hull (now out of the water) and wait until Carol luffed up enough to take pressure off the board and allow me to pull it. Sometimes I looked down at the board cutting the water. I would see a layer of air the whole width and half the immersed depth of the board, bubbling along beside it. Clearly, much of the board was all drag and no lift.

Like sails, leeway preventers must be foil-shaped. Questions abound as to the best chord thickness for a foil in air or water and at various speeds, and also about the best aspect ratio and the best section. Few of us know the answers to those questions, and we couldn't apply them easily if we did, being constricted by the strength of material, class rules, or the depth of water and height of bridges. But anyone who has studied foils soon learns that the leading edge does most of the work, so high aspect ratio is desirable, and the ends of a foil are where power is lost. For best power, a foil should be end-stopped, and this (not the small additional sail area) is why racing genoas sweep the deck.

If a genoa does not sweep the deck, the air traveling across it is not that much denser than the air around it, so the sail still works reasonably well. But a leeway preventer operates in water—835 times denser than air. Except at very slow speeds, as in a sailing kayak, the effect of having no hull to end-stop the foil at the top is catastrophic. To minimize this effect, Dutch leeboards are narrower at the waterline than under water, but the primary way that Dutchmen make their leeboards work is by designing slow hulls.

After the first season, I cut *Monkeys'* bottom and built a daggerboard trunk. To make sure the experiment was equable (if you change more than one thing at a time, you don't know which change caused the improvement), I first tried the leeboard in the daggerboard trunk, with no change in the leeboard's shape or depth. The improvement was unbelievable. She steered much easier, and made much less leeway. I'm certain that no one who has ever tried a good sailboat, first with a leeboard and then with a daggerboard, would ever go back to a leeboard. There simply is no comparison.

Monkeys also started life with a standing lugsail. This 77-square-foot sail was a thrifty part of her design. Several years earlier, we had bought it for the Dobler when we tired of the Sunfish sail. Gilbert Webster made it. Gil may not have been seeing too well that day, but even on his best days, he didn't take much interest in daysailers. It also wasn't ideally designed, because a lugsail should be peaked up more than this one, even though it reduces area. In the last days when standing lugs were raced, yards had become almost vertical, and were easily mistaken for gunter yards. Luffs were short, and did not extend forward of the mast, and were laced to it. By that time, the sail was little different from a low-aspect ratio Bermudan, except that the yard had to be raised and lowered with the sail.

We tried the lugsail a few times in the Dobler, put it in a closet, and made another spritsail. But in the back of my mind was the thought that I hadn't given the sail a fair trial, or fooled enough with its adjustment, and that someday I should. In *Monkeys*, we gave it a very thorough trial, but no matter how it was adjusted, it twisted too much, and would not drive the boat on a close-winded course. The yard could be dipped for each tack with a simple tug on the luff, and this gave an illusion of efficiency. Maddeningly, the part of the sail forward of the mast would backwind while the part aft of it was still filling and drawing. In this mode, and with the leeboard gurgling quietly about its own business, we would point to windward and slide sideways across the river, making what we could on the tide. Downwind, the lugsail did better, though it still took more twist than other low sails, and like a lateen sail, it was always too full in strong winds and too flat in light ones.

Before very long, we converted the sail to gaff by cutting off the area forward of the mast. This reduced area 12 percent, and made an incredibly faster boat. It was certainly as much of an improvement as going from leeboard to daggerboard. It was after these two changes that *Monkeys* became a pleasure to sail. The gaff rig's center of effort is well aft of where the lug center was, but that has never been a problem, and has made no discernible difference. The sail still isn't a good shape (pace, Gilbert!), and the years haven't improved it. But with the gaff sail, the boat is faster on every point, except dead downwind in drifting conditions. She is very manageable for a hull with less than a 2.5 length-to-beam ratio. Sitting in the bottom, she gives the feel of a much bigger boat. I've had a barrel of fun with *Monkeys*, and sometimes when I can borrow her for an afternoon, I still do.

El Toro

Barrel of Monkeys was sold because I was fascinated with garveys, and thought I could improve on Chris' boat. The garvey was finally sold because both Carol and I wanted to try one-design racing in one-man boats. The smallest one-design that is actively raced by adults in the United States is an 8-foot by 4-foot V-bottom pram, which evolved from a design first published by Charles MacGregor in *The Rudder* in June, 1939. Today MacGregor's name is not heard often, but at that time he, along with Robert M. Steward and William F. Crosby, were writing most of the articles *The Rudder* published about boatbuilding and design.

MacGregor advertised himself as a "plywood authority since 1915," and his column "Plywood for Boats" appeared in every 1939 issue of the magazine. His neolithic ancestor of the Toro had a lug rig, oar steering, lee-

Figure 3–6. Monkeys *with her original leeboard and lug rig.*

boards, and a plywood skin braced with steam-bent frames. He called it a "Sabot," and the various Sabot and Sabotina plans also evolved from it. In the past 35 years, dozens of fiberglass dinghy builders (each seeing his fortune just around the corner, once his labor problems clear up) has stealthily filched MacGregor's lines. It is surely the most copied yacht design of all time.

Not much more can be done to make single sheets of 8-foot plywood carry a load through the water than MacGregor did 52 years ago. His devotion to the material is the more remarkable when we remember how bad plywood was in those days. Chemists were still stumbling toward a waterproof glue, but the earliest really satisfactory glues were technological spin-offs of the second World War. To my knowledge, no pre-war plywood boats are in existence today, and for a fact El Toro #1 was put out to pasture more years ago than anyone now remembers.

The Toros that Carol and I now race against are all fiberglass, and many of our competitors never saw a wooden Toro until we appeared with two of them. On the West Coast, where Toros race in larger and more competitive fleets, wooden Toros are still being built, and they often win the Nationals. Just as you can't build a good steel boat under 40 feet, or a good welded aluminum one under 30 feet, you can't build a good fiberglass one much under 15 feet; by the time each square foot of skin is stiff enough for the load, it's too heavy. The Sunfish, for example, is still doggedly built in fiberglass by the thousands, in imitation of the handful of 1/4-inch plywood ones built 40 years ago. Its shape is not inherently stiff to begin with (the way a round bottom would be), and it has a skin so thin that it oilcans everywhere; the slightest grounding is likely to carve a centerboard trunk where the daggerboard trunk used to be. For the mass producer of small boats, polypropylene over foam billets is certainly a better material than fiberglass, but the tooling is more expensive.

In the two years since I built our El Toros, class rules have changed dramatically, and the old building method is now only historically interesting. It required that wooden boats be built on a jig, and specified nearly every piece of lumber in them. Now, wooden Toros can be built in any manner, so long as they measure correctly. Suddenly, taped seams, flotation compartments, and transom bailers become possible. Our boats have been sold, and I look forward to starting on new ones.

Any man who hopes to interest his wife in sailing should get her a boat of her own, small enough to be singlehanded. He should get another for himself, so that he will be too busy to stand on the dock and bawl instructions. Many men would rather give up sailing than do it that way, especially—as has happened in our case—if the wife turns out to be a better dinghy sailor than her husband.

It was on her own Toro, not in the Dobler, *Monkeys,* or the garvey, that Carol really learned to day sail. She loves it; she is good at it; and she is still improving. Her lighter weight is an advantage in most conditions, but she also concentrates better than I do, and moves around quicker and easier in the boat. When she tacks the boat, she tacks her own body too, but I have to jibe my long legs around my bum, losing sight of the course, and sail while I do it. On all courses, she relies on a masthead windvane,

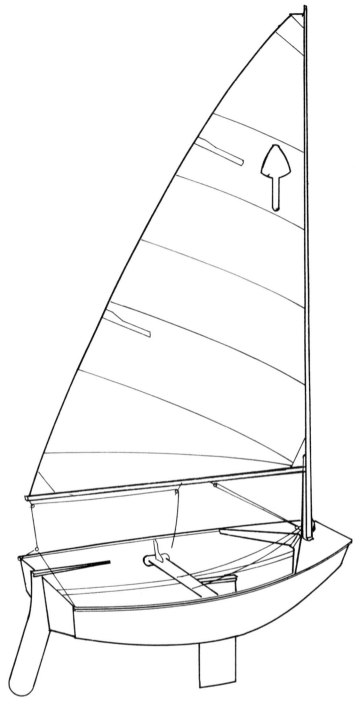

El Toro one-design racing pram.

and it fools her less often than my sail yarns and the hair on the back of my neck. Locally, the class is dominated by three 15-year-olds who weigh about what Carol does and move around their boats like gymnasts. She'll never be as agile as they are, but they won't always be as light as she is either.

El Toros are far from perfect boats, and that's true of most one-designs. The whole idea of a one-design is that you sail in a boat less good than the best you can imagine, in order to race on equal terms with other, identical boats. This is a profoundly anti-capitalist idea. Most of us expect to buy a better boat, and win that way. The one-design seems almost demeaning. Periodically, development classes are established, to give us room to swing our wallets. The 12-Meter Class is probably the most famous, though there have been dozens of others. The Moth is easiest to discuss. Originally, the only specifications were 11 feet overall, and 90 square feet of sail.

For a time, the Moth class was active in many areas, though everywhere it evolved steadily toward a single sail with increasingly complex controls, and a plumb bow and transom. A beam restriction had to be added, because Moth catamarans were blowing the fleets away. The last development was a very narrow single hull, with permanently fixed hiking wings that brought the hull out to maximum beam. Naturally, these boats had to be sailed flat, because dipping a wing was ruinous to speed and control. For all their speed, they were more work than fun to sail, but in the right conditions, no other Moth could touch them.

There the Moth class sank. Long ago, 12-Meters would have sunk too, if not supported by the trimaran-like floats of jingoism and real estate promotion. Development classes inevitably move to extremes or to obsolescence. And in the end, they make one-design classes look inviting again.

When the El Toro class was established, MacGregor's hull lines were changed very little. Leeway prevention came inboard; the steering mechanism was attached to the transom; and the rig went Bermudan, 10 feet 11 inches at the luff and 7 feet across the foot. There the class was frozen. Over the years, restrictions were placed on sail attachment to spars, on sail controls, and even on the gear that had to be carried (a bailer, a paddle "at least as large as a ping pong paddle", a 15-foot painter). The one thing the class association could not freeze was the evolution in all sailboats toward getting more lift and less drag from each square foot of sail. Roach was limited only by number and length of battens, so it was quickly optimized. Dacron sailcloth, which arrived in the '50s, held its shape better than canvas, and was lighter and less porous. A great, unmeasured bag appeared in the foot of the sail, helping to end-stop it. The luff grew baggy

Figure 3–7. El Toro sails are cut very full.

too, and was controlled by bending the mast. The sailplan, originally intended to be 37 square feet, is now more like 50, and each square foot drives the boat with more force.

Hank Jotz, several-time national champion and our own sailmaker, insists that we put the black bands of our masts on sawhorses and hang a 50-pound weight between them. If the bend is less than 2 inches, he wants the number, and he cuts the sail to suit. If it's more, he won't make the sail. He uses mostly 3-ounce Dacron, so hard-handed that when luffing, it sounds like strings of firecrackers going off. In the baggy foot, he uses some bits of what looks like rip-stop nylon. He doesn't tell you any more than you need to know about what he's doing. Other sailmakers favor Mylar Toro sails, but Hank seems on top of the technology, and he still favors Dacron.

Hank also favors masts that bend only in the top 4 feet, and so do most racers. Such masts are heavy, and in certain conditions (especially when jibing) can steer the boat more effectively than the rudder, though not as often in the right direction. Class rules allow Cunninghams, but Jotz doesn't favor them. Instead, his sails are arranged so that the tack floats free, and is controlled by a combination downhaul/inhaul that tensions both luff and foot. To use this effectively, mast and boom tracks need to be cut away to the maximum allowable 15 inches. Mast rake is also important to Hank. Class rules allow some adjustment of it, though not while racing.

All this stuff is pushing on that little cigar-box of a hull designed in 1939, so sailing one is something like driving a fuel funnycar over a goat track in the mountains. You may be wet in an El Toro; you may be cramped and sore; you certainly aren't going very fast; but you aren't bored. Toros can capsize any number of ways—from putting the lee rail under when jibing, or in a gust upwind. The windward rail can also go under during rhythmic rolling downwind, especially if there is water in the bilge. Toros can submarine—come down a wave and keep on going down. I've never seen one capsize transom-first, but it wouldn't surprise me.

To tack in strong winds, it is often necessary to raise the daggerboard while coming through the eye, and hold it up until the boat is moving well on the new tack. If you don't, she can get into irons and stay there, coming off on first one tack and then the other, but heading up again before you have steerageway.

These boats give new meaning to the word *crank*, but I doubt that other one-designs that were designed to build easy in wood, rather than to move easy through the water (e.g., GP-14, Snipe, Star) are any less crank, with the vastly more powerful rigs they now have. Forty years ago, I raced Lightnings off the Maine coast, where afternoon could bring a fresh wind and steep seas, but the boats weren't too troublesome. Recently, I daysailed a friend's modern Lightning, and found it a wholly different boat. The Dacron sailcloth, the multiplicity of sail controls, and the steeply raked and bendy mast overpowered the fat, slab-sided hull, just like a Toro. If you want a one-design that is both fast and manageable, best get one that was designed for a modern rig, like a Laser.

For all that, El Toros are handy boats. Two of them will fit on top of a small car. Minimum weight is 80 pounds, and it is generally agreed that the rig, rigging, and daggerboard will weigh about 22 pounds, leaving 58 pounds for the hull. Fiberglass boats are usually heavier than that, but wooden ones needn't be. When not in use, Toros can be stored in very small spaces, even vertically in closets. They are an awkward shape to carry, however, so when going any distance, we use a wheel on a stick that fits into the mast step.

Figure 3–8. Trundling wheel.

Toros have their own charm too. They are so tiny that sailing them is a challenge, and in light conditions, they can be delightful. Ghosting with the sheet eased, they have no trouble staying with as good a boat as an *Albacore*. They turn in even smaller creeks than *Monkeys* does, and probably could be sailed in an average-size swimming pool. Each spring, a Toro regatta is held in the Tidal Basin at Washington, D.C. The basin is 100 acres, but there is room for dozens of Toros and all kinds of racing tactics. On the other hand, Toro fleets like to sail in big water: until recently they raced across Chesapeake Bay every year. They still do race across San Francisco Bay, with one crash boat for every two competitors.

The rig of a Toro is very sensitive to adjustment, and slight changes in the controls can make dramatic differences in the speed. It's interesting to study. The boats do slam into waves, and can even slam to a standstill, although with an alert and sensitive helmsman they slam less often and less hard. Because crew can easily be two-thirds of total weight, crew position is crucial and ever changing, which is why the gymnast kids do so well.

Toros make the best of their name: The class emblem is a dung shovel, and the class newsletter is the *Shovel Bulletin*. In discussing boats and races with other skippers, bragging and lying are always well received— it's bull shooting. Today, many of the East Coast Toro skippers are old enough to be retired. Some have plastic hips or hearing aids. In their

Figure 3–9. Toros racing in the Tidal Basin, Washington, D.C.

boats, they tend to mutter to themselves and sort through their gear like a mouse in a cage. But they have been racing these boats 20 years or more, and they usually go past you. When they do, they will often offer a suggestion about sail or boat trim, which can be confusing to new skippers who are most used to the cut-throat attitude of other American sports. Carol and I like it, and we especially like the two-day regattas when the whole class goes out to dinner together. We have no plans to move to another class.

El Toros are not simple to rig. Because the sail must slot into boom and mast, it must be taken off the spars after each use. Haul, sheet, and vang take some time to attach, and many skippers use another line called a "J.C. strap" which holds the boom out and down in light air downwind. Even without a J.C., it takes us 15 minutes to rig the boat, and many trips back and forth to the shed. Before we had owned our Toros a year, I was thinking about another daysailer, not for racing but just for sailing on the Tuckahoe River.

Tuckahoe Ten

The new boat was a kind of anti-Toro, and was meant to be comfortable, simple to rig and sail, at least as fast as a Toro, and not in any circumstances crank. Sailing a kayak had always interested me, despite the slow speeds they made. A short, fat kayak could carry more sail, could tack, and could still be paddled when necessary. The skipper could sit as in a kayak, facing forward with a backrest behind him and his legs out in front. I had never seen such a boat, but the numbers suggested it could work, at least in light, inland winds.

As the boat was primarily for sailing, the hull shape should be more like a sailing dinghy, even if the proportions were more like a kayak. Tom Colvin says that in considering the design of a boat, "The primary use is primary. All other uses are secondary." This may sound simplistic, but it is the hardest of all truths to keep in mind.

Hull lines for sailing dinghies have been well worked out over the years. The great breakthrough of this century was Uffa Fox's invention of the planing dinghy in 1928. But however much I admired his boat, I didn't want a planing kayak. I looked instead to Howard Chapelle, a very sound middle-of-the-road designer, and specifically to his 10-foot sailing dinghy (*Boatbuilding*, p. 421). This boat, and Nat Herreshoff's 11½-foot *Columbia* tender, are two of the best small boats ever designed. Both have excellent lines forward, though I prefer Chapelle's with less hollow in the waterlines. Aft, the Chapelle has a harder bilge, less deadrise, and a flatter run. The Herreshoff must row better, but the Chapelle is the choice for sailing. Both boats have Bermudan rigs with luff and foot the same length, and this may be as bad a rig as the standing lug. But with Chapelle's lines beside me, I began drawing the Tuckahoe Ten.

Ten feet overall, and as much of that as possible in the water, seemed a good way to limit wetted surface and hull weight. Reducing the beam by 25 percent left the widest boat that could be paddled, and the narrowest that could carry useful sail without a sliding seat. Reducing hull depths by the same 25 percent brought displacement to 234 pounds, which seemed reasonable for boat, rig, and skipper. Sheer height was then reduced a bit more, especially at the bow where Chapelle had become a trifle grandiose. Every square inch of skin weighs something. Final dimensions were 10 feet 2 inches by 3 feet 1 inch, and just under a 5-inch draft.

The next consideration was building material and method. Chapelle calls for Ashcroft planking: two layers of ⁵/₃₂-inch veneer, diagonally laid, with plenty of stringers, brads, and glop. Ashcroft was a predecessor of cold molding, before the days of waterproof glue. Today, I could just cold

Figure 3–10. Tuckahoe Ten under sail.

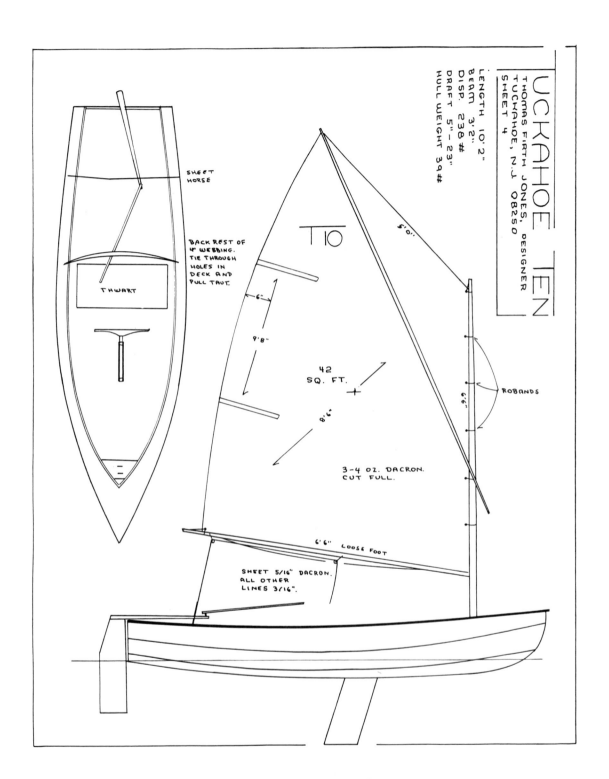

TUCKAHOE TEN
THOMAS FIRTH JONES, DESIGNER
TUCKAHOE, N.J. 08250
SHEET 4

LENGTH 10' 2"
BEAM 3' 2"
DISP. 238 #
DRAFT 5" - 23"
HULL WEIGHT 39 #

SHEET HORSE

BACK REST OF 4" WEBBING. TIE THROUGH HOLES IN DECK AND PULL TAUT.

THWART

T 10

5' 0"

6"

9' 8"

42 SQ. FT.

8' 6"

9' 6"

ROBANDS

3-4 OZ. DACRON. CUT FULL.

6' 6" LOOSE FOOT

SHEET 5/16" DACRON. ALL OTHER LINES 3/16".

92

OFFSETS IN INCHES AND EIGHTHS TO INSIDE OF PLANK. STATION SPACING 14 1/2".

Bow	HALF BREDTHS SHEER	UPPER	LOWER	TO BASE SHEER	UPPER	LOWER	KEEL
1	9-4	7-5	4-1	2-4	6-4	13-4	19-2
2	14-4	12-4	7-4	4-2	9-7	16-2	20-4
3	17-2	15-4	10-4	5-2	12-1	17-6	20-5
4	18-2	17-4	12-6	6-1	13-3	18-4	20-5
5	18-3	18-0	13-2	6-4	13-5	18-6	20-5
6	17-5	17-2	12-7	6-3	13-5	18-5	20-3
7	16-0	15-5	11-5	6-2	12-6	17-6	19-4
TRAN	14-0	13-5	10-0	5-3	9-7	14-0	16-0

STEM TO BASE: STA. 0, 14-3. STA. 1/2, 17-7.
STEM FWD. STA. 1: SHEER 18-6, UPPER CHINE 18-2,
LOWER CHINE 15-4, W.L. 12-2.

TUCKAHOE TEN
THOMAS FIRTH JONES, DESIGNER
TUCKAHOE, N.J. 08250
SHEET 1

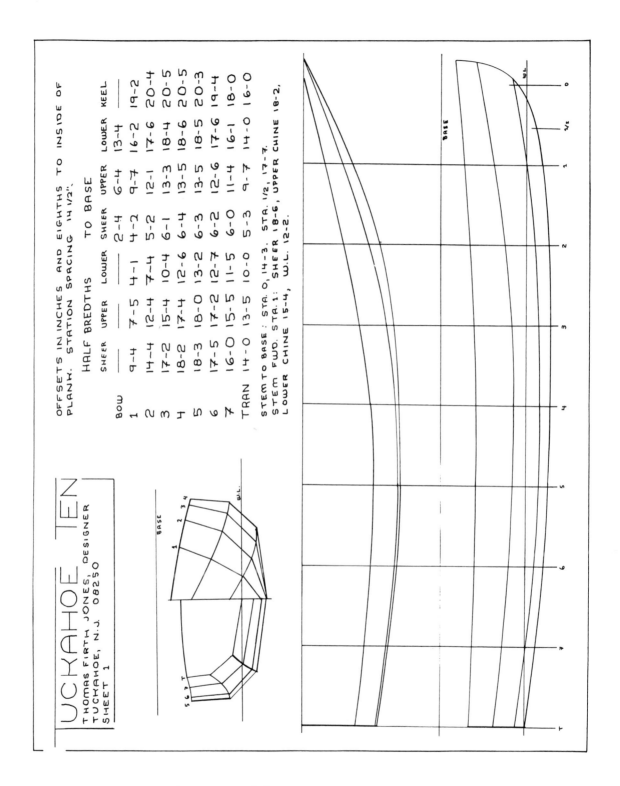

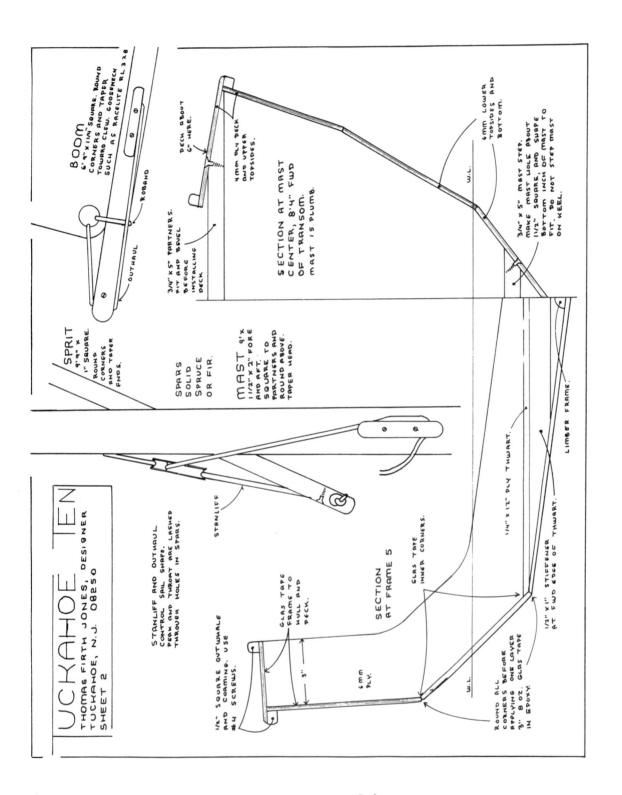

BOOM

4'4" x 1" SQUARE. ROUND
CORNERS AND TAPER
TOWARD CLEW. GOOSENECK
SUCH AS RACELITE RL328

ROBAND

OUTHAUL

SPRIT
4'4" x
1" SQUARE.
ROUND
CORNERS
AND TAPER
ENDS.

3/4" x 5" PARTNERS.
FIT AND BEVEL
BEFORE
INSTALLING
DECK

DECK ABOUT
6" HERE.

4MM PLY DECK
AND UPPER
TOPSIDES.

SECTION AT MAST
CENTER, 8'4" FWD
OF TRANSOM.
MAST IS PLUMB.

6MM LOWER
TOPSIDES AND
BOTTOM.

W.L.

3/4" x 5" MAST STEP.
MAKE MAST HOLE ABOUT
1 1/2" SQUARE, AND SHAPE
BOTTOM INCH OF MAST TO
FIT. DO NOT STEP MAST
ON KEEL.

SPARS
SOLID
SPRUCE
OR FIR.

MAST 9' x
1 1/2" x 2" FORE
AND AFT.
SQUARE TO
PARTNERS AND
ROUND ABOVE
TAPER HEAD.

TUCKAHOE TEN

THOMAS FIRTH JONES, DESIGNER
TUCKAHOE, N.J. 08250
SHEET 2

STANLIFF AND OUTHAUL
CONTROL SAIL SHAPE.
PEAK AND THROAT ARE LASHED
THROUGH HOLES IN SPARS.

STANLIFF

1" SQUARE OUTWHALE
AND COAMING. USE
#4 SCREWS.

GLAS TAPE
FRAME TO
HULL AND
DECK.

SECTION
AT FRAME 5

GLAS TAPE
INNER CORNERS.

1 1/4" x 12" PLY THWART.

LIMBER FRAME

1/2" x 1" STIFFENER
AT FWD EDGE OF THWART.

3"

6 MM
PLY.

W.L.

ROUND ALL
CORNERS BEFORE
APPLYING ONE LAYER
3" 8 OZ. GLAS TAPE
IN EPOXY.

94

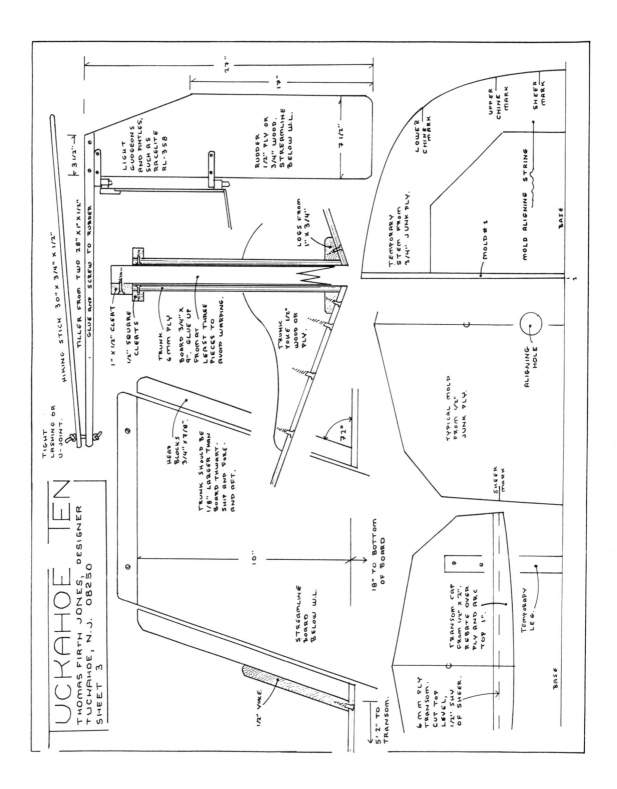

TUCKAHOE TEN

THOMAS FIRTH JONES, DESIGNER
TUCKAHOE, N.J. 08250
SHEET 3

TIGHT LASHING OR U-JOINT.

HIKING STICK 30" x 3/4" x 1/2"

TILLER FROM TWO 28" x 1" x 1/2"
GLUE AND SCREW TO RUDDER.

├ 3 1/2" ┤

LIGHT GUDGEONS AND PINTLES, SUCH AS RACELITE RL-358

27"

17"

RUDDER 1/2" PLY OR 3/4" WOOD. STREAMLINE BELOW W.L.

7 1/2"

1" x 1/2" CLEAT

1/2" SQUARE CLEATS

TRUNK 6MM PLY

BOARD 3/4" x 9". GLUE UP FROM AT LEAST THREE PIECES TO AVOID WARPING.

LOGS FROM 1" x 3/4"

TRUNK YOKE 1/2" WOOD OR PLY.

HEAD BLOCKS 3/4" x 7/8"

TRUNK SHOULD BE 1/8" LARGER THAN BOARD THWART-SHIP AND FORE-AND AFT.

72°

STREAMLINE BOARD BELOW W.L.

10"

18" TO BOTTOM OF BOARD

1/2" YOKE.

TEMPORARY STEM FROM 3/4" JUNK PLY.

UPPER CHINE MARK

SHEER MARK

LOWER CHINE MARK

MOLD ALIGNING STRING

MOLD #1

BASE

TYPICAL MOLD FROM 1/2" JUNK PLY.

ALIGNING HOLE

SHEER MARK

6MM PLY TRANSOM. CUT TOP LEVEL, 1/2" SHY OF SHEER.

TRANSOM CAP FROM 1/2" x 2". REBATE OVER PLY AND ARC TOP 1".

TEMPORARY LEG.

BASE

5'2" TO TRANSOM.

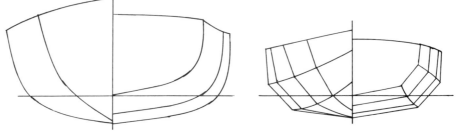

Figure 3–11. Chapelle's 10-foot dinghy (left) and Tuckahoe Ten (right).

mold, but even a 10-foot dinghy is more cold molding than I ever want to do.

Strip planking was a real possibility. Conventional strip plank, with each strip nailed and glued to the next, can't be much less than ¾-inch thick, which is too heavy for a kayak. But there is a kind of strip building, popular with canoe builders 20 years ago, where cedar strips about ¼ inch by ¾ inch are bent over many molds, and just stapled to each other until the glue cures. Then the staples are pulled and the hull is sanded and glassed outside and in, making the strips a core material. I was attracted to this method, especially as I had never tried it, but at last put it aside because I feared that it would be too much like cold molding—hundreds of fiddly bits misaligning themselves despite my glue-sodden imprecations. In addition, glued softwood structures are the devil to sand, because the glue is harder than the wood and stands proud while the wood is sanded away.

Various other materials—fiberglass, clinker, fabric over stringers—were considered and discarded for various reasons: weight, wetted surface, fragility. At last I came down to the familiar taped-seam plywood. For this boat, I used 6-millimeter 5-ply okume for bottom and intermediate planks, and 4-millimeter 3-ply for upper planks and deck. Taping the seams was chosen over framing, not because it is easier, but because it is lighter and it allows more rounding of the chines. In fact, a framed structure would be easier to build in a hull with this many chines.

Conical Projection

To convert Chapelle's round sections to the chined sections suitable for plywood, the panels with much twist were checked to see if they would fit into the surface of a cone. This is called conical projection. The drawing shows how it was done for the forward half of the bottom (Figure 3–12). An apex for the cone was chosen, though a drawing is seldom completed without moving the apex a number of times. The best place to

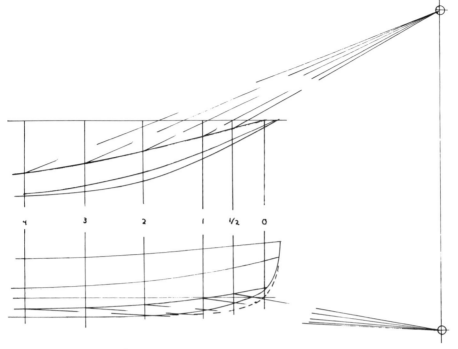

Figure 3–12. Conical projection.

start is with someone else's projection drawing that shows a boat intended for similar service, or with a similar prismatic coefficient.

The two circles in the drawing represent the *same* apex of a single
cone, but one shows it from above (plan view) and the other from the side
(profile). Naturally, the two representations of the apex fall on the same
vertical line. From this apex, generators radiate out to each point where
the chine crosses a station. The generators are lines drawn on the surface
of the hypothetical cone. The drawing shows two representations of each
generator, one in plan and one in profile.

When vertical lines (omitted in the drawing) are dropped from where
the generators cross the keel in plan (that is, the centerline) down to the
generators in profile, the keel line in profile *must* go through those
points. If it doesn't, the skin of that bottom panel will not fit into the surface of a cone, and it can't be planked in sheet goods.

Choosing another spot for the apex will often yield a shape closer to
the original lines drawing, but usually the lines have to be altered somewhat to suit the tyrannical shape of the cone. Sometimes the chine can
be fiddled one way or the other, but usually the keel profile has to be altered as well. The dotted line shows the keel and stem profile that I

hoped to achieve in the Tuckahoe Ten. The solid one shows what I settled for.

This type of drafting must be very accurate, and must be done on paper that will stand much erasing. Using the same generators, any number of waterlines and buttocks can be drawn, which will show the curves (or "development," as it is called) of the sections. This is necessary on plywood boats where the frames will be left in, but not necessary for the Ten. It is worth remembering, however, that conical projection is a two-stage process: The first to find a shape that is a conical surface, and the second (if needed) to find the sectional curves on that surface.

The intermediate planks of this boat also twist some, and theoretically they should be developed, too. But the twist is slight compared to the bottom forward, and plywood can be unnaturally twisted a certain amount if need be. As explained in Chapter 1, this torturing is done by experiment, not by drafting. In the Ten, I got away with torturing the intermediates.

The hull was built somewhat differently from other taped-seam boats. I used two permanent frames (transom and #5), six molds, and a temporary stem. Juggling six panels together, it would be tough to cut them out by the numbers, tack them together, and expect a fair or symmetrical shape to result. My planks were fitted as if to a framed hull, and because the ply was expensive, I made patterns (and mistakes) with 1/8-inch Masonite, before cutting the ply itself. I used as little wire as possible, often driving temporary screws into the molds instead.

Figure 3–13. Tuckahoe Ten molds and transom.

Figure 3–14. Planking pattern made from 18-inch Masonite.

I used epoxy resin and 8-ounce glass cloth with very little binder. The binder in glass (especially thick in mat) holds the strands together until it is dissolved by the styrene in polyester resin. But there is no styrene in epoxy, so the binder remains on the glass strands, and the epoxy encapsulates them, but does not adhere to them. I cut my own tape (about 2 inches) from a roll of cloth, because manufactured tape has selvage edges that stand up and are dense and hard to sand. Both these things—the lack of binder in the cloth, and the cut edges of the tape—mean that many glass strands tangle with brush and gloves, and make a pretty good mess. But these wicked-up strands are very easy to sand off, once the resin is cured.

By the way, I buy rubber gloves for about $15 a hundred, and use them for even the most minor epoxy jobs. I wipe the gloves and tools off with paper towels, and avoid acetone as much as possible. If I have a big epoxy job to do, I save it until the end of the day, so I'm out of the shop while it's curing. I wear a respirator for sanding. Like booze, epoxy is wonderful stuff, but it don't do you no good.

When the Ten's hull came off the molds, it was surprisingly rigid and gratifyingly light. It still needed skeg, decks, seat, trunk, and mast step

Figure 3–15. Tuckahoe Ten hull, with optional skeg.

and partners. The skeg is unnecessary drag when sailing, but though the boat is narrow for sailing (32 inches on the water) it's still beamy for a paddling boat, and lacks the deadrise at the stern that keeps a paddling boat running straight. To minimize drag, I kept the skeg small, and gave it a foil section like a centerboard.

Rudder and tiller are glued together. As the forward end of the tiller is nearly under the sheet horse, it can't be raised much anyway, so I thought to eliminate slop from one joint. The hiking stick has a good universal joint with very little slop. All this gear is very light—the hiking stick is only ½ inch by ¾ inch. This gives a feeling of delicacy and precision, as well as saving weight.

The seat is for comfort, of course, and also to make shifting my weight quicker and easier. Lacking one of Rushton's patented centerboards, I'm stuck with a board trunk coming up between my legs. Lateral support for it is a wooden yoke not much higher than the seat that comes up under my knees, its edges well rounded. The trunk is low enough to get a leg over when need be. Compared to other daysailers its size, the Tuckahoe Ten is almost luxuriously comfortable to sail, especially when the wind is light or astern, and you'd be squatting in the middle of a pram.

Sailing Rig

In a boat this narrow and tender, there is little merit in the Bermudan rig. At the dock, if you stuck a Toro mast in her, the boat would have to be held upright until you were seated in her or she'd capsize from mast weight alone. Some designers solve this problem with a very light, small-diameter mast, and usually it does not break. But we had three sets of masts in our Toros before making the right ones, and we knew that although a bendy mast can control sail shape, too bendy is worse than too rigid. It is useless in any but the lightest air, because it takes all the shape out of the sail.

The Ten has a spritsail, a very good one, from Hank Jotz. It's like an Optimist sail, but has 6 inches more luff and 12 inches more head. It is loose-footed, and has a big foot roach to make an end-stopping shelf. It

Daysailers

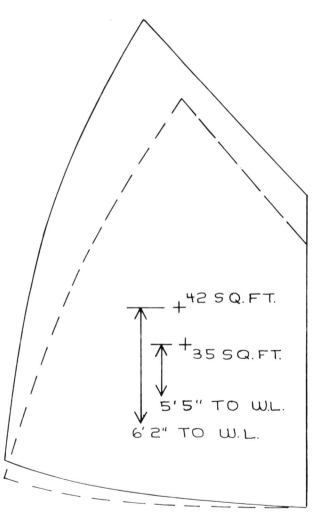

42 S Q. FT.

35 S Q. FT.

5' 5" TO W.L.

6' 2" TO W.L.

Figure 3–16. Optimist (dashed lines) and Tuckahoe Ten sails.

Figure 3–17. The Tuckahoe Ten wearing her winter Optimist pram rig.

has battens and a stanliff. Boom and all, it can be brailed to the mast. We also have an old Optimist sail for the Ten, using the same mast and boom but a shorter sprit. This is kept on the spars in winter, when winds are gustier and the water is cold.

The boat is indeed simple to bring to the water and rig. The hull weighs 39 pounds. One more trip to the shed fetches the rest of the gear, which weighs another 15 pounds. Because there are no lines for sail adjustment (sprit rig needs them less than Bermudan), I can be under way in a minute or two. After the mast is stepped, I just release the brail line, thread the sheet through the boom blocks, and sail away.

If the rig had to be taken apart each time the boat was used, the brail would be more trouble than it was worth. Even without it, setting the rig up would be at least as much trouble as a Toro. But if the boat can be kept near the water, there's almost nothing to it. Like a Toro sail, the Ten sail would hold its shape longest if taken off the spars after each use, rolled up, and slipped into a bag without folding. Jotz believes in this so firmly that his dinghy sails come with a bag longer than the foot, and no doubt it makes a difference in one-design competition.

There isn't any Tuckahoe Ten racing, and the Ten doesn't give the skipper enough to do to be an interesting racer. The sheet is the only line to

Figure 3–18. Carrying the 39-pound Ten.

pull. The seat and backrest preclude running around in the boat. Racing and recreational sailing should both be lively, but I like my recreation simple, and my racing complex.

Striking or setting the rig under way, the board trunk is an encumbrance, but it's still possible to do. For paddling, the brailed rig will stow with the mast heel forward and a wad of sail and spars under the skipper's armpit. But the Ten is just a marginal boat under paddle, and the paddle must come apart, or else it's in the way on one tack or the other. I use a 9-foot military surplus liferaft paddle of tubular aluminum. For good paddling, the Ten is too high-sided, as well as too short and wide and flat-transomed.

We really enjoy sailing the Ten alongside a Toro, here on the river. They're a close enough match so that Carol usually stays ahead of me, whichever boat she's in. Sometimes we stop at a neighbor's dock and change boats, and it's always a surprise, because they feel so different to sail. That's because the Toro gets her performance from a powerful rig and long righting arm, and the Ten gets hers from low resistance.

Upwind in a gust, the Ten skipper has to luff, while the Toro skipper

Figure 3–19. The Ten under her brailed rig.

can shift his weight to windward and keep the sail full. But Ten, with her 3-foot-longer waterline, is still footing as fast as the Toro, and pointing as high. The bow half-angles are 15 degrees and 47 degrees; length-to-beam ratios are 3.6 and 1.9. Off the wind, it would seem that the Ten should walk away from the Toro. But thanks to her long waterline, she does have a bit more wetted surface. She probably has 10 percent less actual sail area, counting all roaches and pockets, and because the sail is lower, it is more often blanketed by trees or even marsh grass. Again, the two boats are so closely matched that if one skipper pulls the daggerboard up, the other had better too.

The big difference is that the Toro skipper is working like a navvy, and the Ten skipper is just sitting there. For him, everything seems in the right proportions and the right place. Seated facing forward, he doesn't

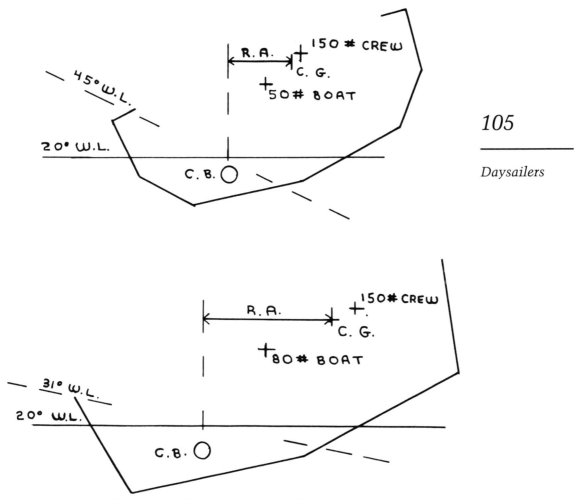

Figure 3–20. Stability of Tuckahoe Ten (top) and El Toro (bottom).

have to jibe himself when he jibes the boat. The sheet, traveling along the boom, comes down conveniently to hand. After a dozen tacks, it hasn't become a giant snarl in his crotch. To shift his weight, he puts his elbows on the side decks and shifts his butt to the other side. Well heeled, he must sometimes put his leeward foot against the topsides to hold himself up to windward; but within the limits of her stability, the Ten is entirely mannerly. She never misses stays or gets in irons. Nothing makes her head up or off except pressure on the tiller. At the end of a several-hour sail, her skipper isn't at all stiff or sore. The Toro skipper sure is.

Downwind in strong conditions, both boats will develop rhythmic roll at the same alarming rate. This problem cannot be solved in monohull

105

Daysailers

sailboats. However, the Ten rig can be scandalized in a few seconds, and the boat can roll much further than the Toro without shipping water. The narrow beam helps with that as much as the side decks. Decks, beam, and hull shape make the Ten about as dry as a 10-foot sailboat can be. I sailed her with the fleet at a recent Toro regatta, when the wind was fresh. After each heat—and sometimes during a heat—Toro skippers had to abandon sailing and wield the bailer. I had only a sponge in the Tucka-hoe Ten, and never had to use it.

Just the same, you can't help thinking what fun you could have with a low-resistance hull *and* a powerful rig. We still have our original Toro masts, which proved too light and flexible, even in moderate winds. Someday I'll fit one to the Ten, and then when the wind is light and air and water are warm, I'll put on a bathing suit and. . . .

4. Long, Narrow Powerboats

Many lakes in state parks allow no powerboats, except perhaps for electric outboards, so the paddler or sailor can escape entirely from the noise and stink of the land once he's away from the launching ramp. This certainly is pleasant, and more lakes could profit from such regulation. But such lakes usually are small, and even so it will be noticed that the ranger usually does have a powerboat, as befits his dignity and the urgency of his function. Usually it's a big, noisy outboard, in a hull throwing the maximum wake.

Nowadays, sculling usually is coached from powerboats as bad as the ranger's, but in the early days of such coaching, considerable thought was given to designing coaching launches half the length of the shells. They would go faster than the shells, but without making enough wake to disturb them. Weston Farmer showed us in *From My Old Boat Shop* some of these low-resistance boats. They had just enough beam to keep careless crew from capsizing them. This kind of hull still makes the best powerboat for sheltered water. It make relatively little noise and stink by burning very little fuel for its speed.

In buying and owning a boat, you pay for volume, not length. Given the same freeboard, a boat 21 feet by 5 feet costs no more than a boat 15 feet by 7 feet. It will be no heavier, and will be just as easy to trailer, paint, and store. It will burn a quarter of the fuel. But you will be hard put to find one in today's market.

Puxe

Puxe was built eleven years ago. She cost 400 hours and $500, not counting the engine. Though lines were drawn on a drafting board, she was built by cleating the topsides together on the floor, screwing them to the stem, and bending them around a mold or two. When I came to draw stock plans for the boat, however, I was less enamored of that method; it really doesn't save much time, and can result in a hull that is asymmetrical or that differs substantially from the intended shape. The plans show the more conventional frames-on-a-strongback method. Despite that, the original *Puxe* has been an excellent boat, rigid and tight, though she has no chine log.

Like the sailing garvey, *Puxe* uses parallel-sided planks in the topsides,

Figure 4–1. At 1.7 times her hull speed, Puxe *is not planing and is easy to control despite a tidal rip.*

Figure 4–2. Cable steering really needs only one spring.

Figure 4–3. Small duckboard postpones gas tank rusting.

and it's as easy to scarf them up on a bench as to butt block them on the boat. In the plans, they are bent around seven mahogany frames, stem and transom. Sheer clamps and chine logs are notched in, and the keelson is laminated in place. After planking but before caulking, cleats are worked in between the frames to stiffen the seams. As in all carvel caulking, I use polyurethane compound without cotton or oakum. All brands seem to have air bubbles as they come from the tube. The bubbles should be worked out with a putty knife.

The bottom is cross-strip planked, and although cross stripping a V- or flat-bottom boat is quicker than tapering strips for a round bottom, it still took a quarter of the total hours in building the boat. The only leak *Puxe* ever developed was when one of the strips that included the heart of the tree split open. When you're already discarding sapwood, it's hard to dis-

Figure 4–4. Ten knots with two aboard, and forefoot just out.

Figure 4–5. Puxe *on the Tuckahoe.*

card hearts as well, but it needs to be done. If the heart doesn't eventually split, it will often make a whole board buckle or warp. Wood goes on changing long after it's finished seasoning.

Like all my planked boats, this one is built of Atlantic white cedar, which is so similar to northern white cedar that they are listed together in the *Wood Handbook*. Both are light weight and very rot-resistant, but have annoyingly many knots. The alternative, western red cedar, can still be had in long, clear lengths, though these come from our virgin rain forest in the Pacific Northwest, which we really shouldn't be cutting. Red cedar, like most western woods, splits and spalls easier than eastern woods. Both white cedar and western red will soak up their own weight in water if kept immersed, so *Puxe* is 100 pounds heavier on hauling than on launching day.

The aft three-quarters of the bottom could have been cross planked with boards, and the time saved would have been substantial. Forward, boards wouldn't have taken the twist, and strip planking is quicker than beveling thicker boards and dubbing them off, outside and in. On a bottom of this shape, only the occasional strip needs to be beveled—done with one or two swipes of a block plane. None need be tapered. Nearing the bow, the surfaces that touch chine log and keelson must be beveled on every stick. The resulting surface will be corrugated or stepped, and will need some sanding. This kind of planking with short strips uses wood pretty economically, and a start on the bottom can be made with scraps left from the topsides. But soon enough, whole boards are going to go through the saw, making little sticks. Considering the little maintenance this boat has needed, the perhaps 2000 hours we have used her, and the thousands of hours we plan to use her still, I don't think the 100 hours spent on the bottom were too much.

She was bronze fastened and resorcinol glued. Resorcinol makes just as strong a bond as epoxy if the fits are good. It's made by mixing a maroon liquid with a white powder; the resulting product is Bakelite. Like plastic-resin glue, the powder contains formaldehyde and shouldn't be breathed. Once mixed, resorcinol is not dangerous. It has a longer pot life than most epoxies, and can be thinned with alcohol and cleaned up with alcohol or water until it starts to go off. It stains wood badly, and will bleed through some finishes, like latex paint. It can be bought cheaply in large quantities, although in small quantities is usually no cheaper than epoxy.

Decks were fir marine plywood—an unhappy choice. Although painted white, they checked badly in sunlight after a couple of years. They were sanded, coated with epoxy, and repainted white. Two years later, they had checked again, and this time I glassed them. Fir marine plywood is as strong as any, and as rot-resistant, but it can't stand sunlight. Glassed, it is heavier and more time-consuming than marine mahoganies, and probably just as expensive.

Puxe is a Portuguese word, pronounced *poosh* and meaning *pull*. Our first summer in Portugal, we wasted much time trying to poosh doors open. She was built for a much bigger body of water than we sail in: downriver, we have two 8-foot fixed bridges, which limit sailing but not motoring. Below them is a vast wetland of three rivers and innumerable

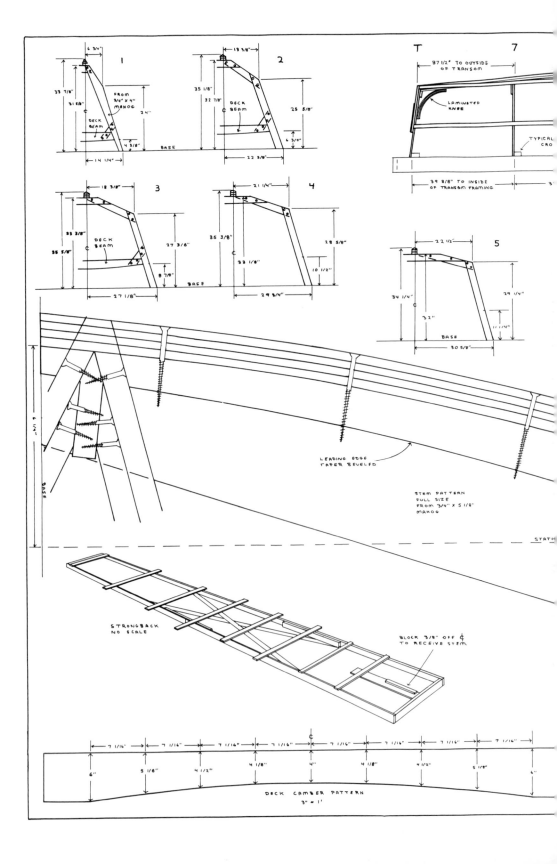

1

6 3/4"

33 7/8"
31 5/8"

FROM
3/8" X 4"
MAHOG

DECK
BEAM

24"

DECK
BEAM

4 3/8"

BASE

14 1/4"

2

13 3/8"

35 1/8"
32 7/8"

DECK
BEAM C

25 5/8"

6 3/8"

22 3/8"

T   7

37 1/2" TO OUTSIDE
OF TRANSOM

LAMINATED
KNEE

TYPICAL
CRO

39 3/8" TO INSIDE
OF TRANSOM FRAMING

3

3

18 3/8"

33 3/8"

35 5/8"

DECK
BEAM C

27 3/8"

8 7/8"

BASE

27 1/8"

4

21 1/4"

35 3/8"

33 1/8"

28 5/8"

10 1/2"

BASE

29 3/4"

5

22 1/2"

34 1/4"

32"

29 1/4"

11 1/4"

BASE

30 5/8"

6 1/4"

BASE

LEADING EDGE
TAPER BEVELED

STEM PATTERN
FULL SIZE
FROM 3/4" X 5 1/8"
MAHOG

STATI

STRONGBACK
NO SCALE

BLOCK 3/8" OFF C
TO RECEIVE STEM.

7 1/16"   7 1/16"   7 1/16"   7 1/16"   7 1/16"   7 1/16"   7 1/16"   7 1/16"

6"   5 1/8"   4 1/2"   4 1/8"   4"   4 1/8"   4 1/2"   5 1/8"   6"

DECK CAMBER PATTERN
3" = 1'

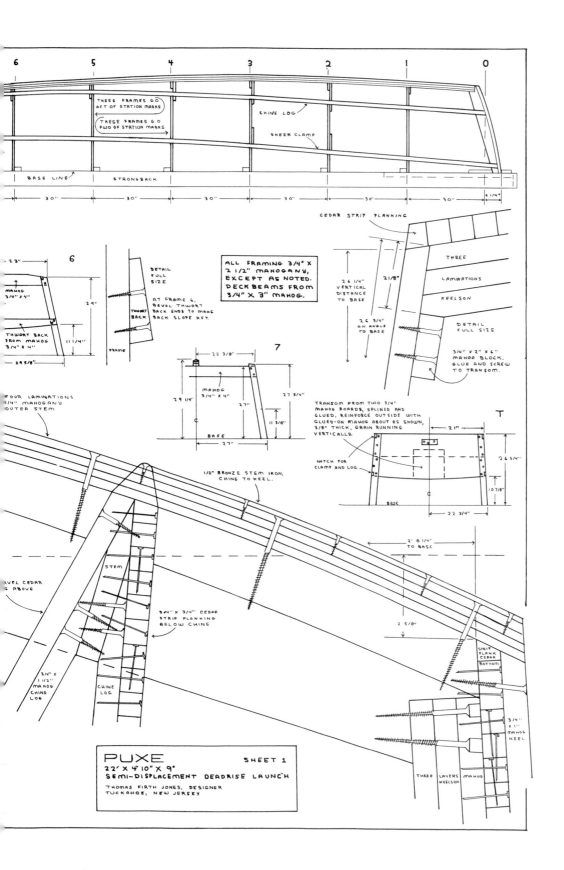

6   5   4   3   2   1   0

THESE FRAMES GO
AFT OF STATION MARKS

THESE FRAMES GO
FWD OF STATION MARKS

CHINE LOG

SHEER CLAMP

BASE LINE

STRONGBACK

30"   30"   30"   30"   30"   30"   6 1/4"

CEDAR STRIP PLANKING

THREE

LAMINATIONS

KEELSON

26 1/4"
VERTICAL
DISTANCE
TO BASE

21/8"

26 3/4"
ON ANGLE
TO BASE

DETAIL
FULL
SIZE

DETAIL
FULL SIZE

3/4" X 2" X 6"
MAHOG BLOCK,
GLUE AND SCREW
TO TRANSOM.

6

- 23"

MAHOG
3/4" X 4"

29"

THWART BACK FROM MAHOG
3/4" X 4"

11 1/4"

29 5/8"

ALL FRAMING 3/4" X
2 1/2" MAHOGANY,
EXCEPT AS NOTED.
DECK BEAMS FROM
3/4" X 3" MAHOG.

DETAIL
FULL
SIZE

AT FRAME 6,
BEVEL THWART
THWART BACK ENDS TO MAKE
BACK   BACK SLOPE AFT.

FRAME

7

22 3/8"

MAHOG
3/4" X 4"

29 1/4"   27"   27 3/4"

11 3/8"

BASE
27"

1/2" BRONZE STEM IRON,
CHINE TO KEEL.

TRANSOM FROM TWO 3/4"
MAHOG BOARDS, SPLINED AND
GLUED. REINFORCE OUTSIDE WITH
GLUED-ON MAHOG ABOUT AS SHOWN,
3/8" THICK, GRAIN RUNNING
VERTICALLY.

NOTCH FOR
CLAMP AND LOG

T

21"

26 3/4"

10 7/8"

BASE

22 3/4"

FOUR LAMINATIONS
1/4" MAHOGANY
OUTER STEM

STEM

2' 8 1/4"
TO BASE

2 5/8"

VEL CEDAR
ABOVE

3/4" X 3/4" CEDAR
STRIP PLANKING
BELOW CHINE

STRIP
PLANK
CEDAR
BOTTOM

3/4" X
1 1/2"
MAHOG
CHINE
LOG

CHINE
LOG

3/4"
X 1"
MAHOG
KEEL

THREE LAYERS MAHOG
KEELSON

PUXE   SHEET 1
22' X 4'10" X 9"
SEMI-DISPLACEMENT DEADRISE LAUNCH
THOMAS FIRTH JONES, DESIGNER
TUCKAHOE, NEW JERSEY

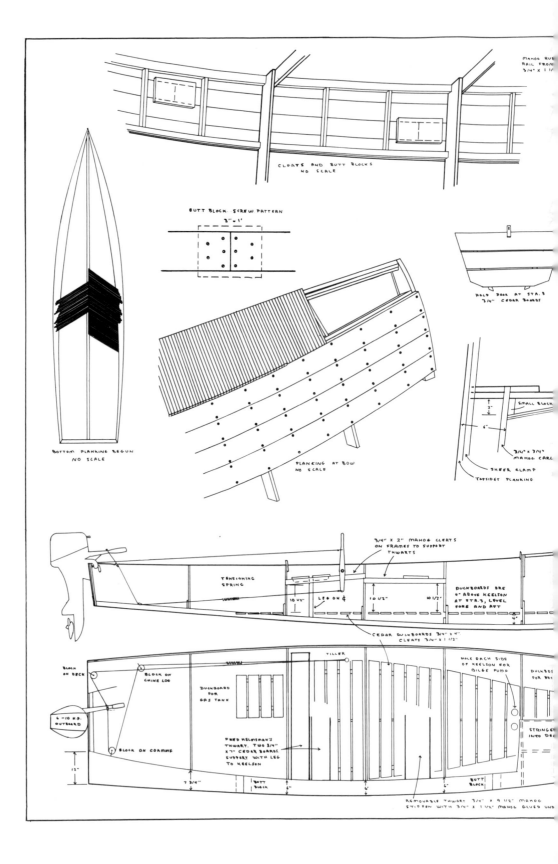

CLEATS AND BUTT BLOCKS
NO SCALE

MAHOG RUB
RAIL FROM
3/4" X 1 1/

BUTT BLOCK SCREW PATTERN
3" = 1'

HOLD DOOR AT STA. 3
3/4" CEDAR BOARDS

BOTTOM PLANKING BEGUN
NO SCALE

PLANKING AT BOW
NO SCALE

SMALL BLOCK

2"

6"

3/4" X 3/4"
MAHOG CARL

SHEER CLAMP

TOPSIDE PLANKING

3/4" X 2" MAHOG CLEATS
ON FRAMES TO SUPPORT
THWARTS

TENSIONING
SPRING

10 1/2"   LEG ON 6   10 1/2"   10 1/2"

DUCKBOARDS ARE
6" ABOVE KEELSON
AT STA. 3, LEVEL
FORE AND AFT

CEDAR DUCKBOARDS 3/4" X 4"
CLEATS 3/4" X 1 1/2"

6 -10 H.P.
OUTBOARD

BLOCK
ON DECK

BLOCK ON
CHINE LOG

DUCKBOARD
FOR GAS TANK

FRED HELMSMAN'S
THWART. TWO 3/4"
X 7" CEDAR BOARDS.
SUPPORT WITH LEG
TO KEELSON

BLOCK ON COAMING

12"

7 3/4"   4"   4"   6"   6"

BUTT
BLOCK

BUTT
BLOCK

TILLER

HOLE EACH SIDE
OF KEELSON FOR
BILGE PUMP

DUCKBO
FOR DRY

STRINGER
INTO DE

REMOVABLE THWART 3/4" X 9 1/2" MAHOG
STIFFEN WITH 3/4" X 1 1/2" MAHOG GLUED UND

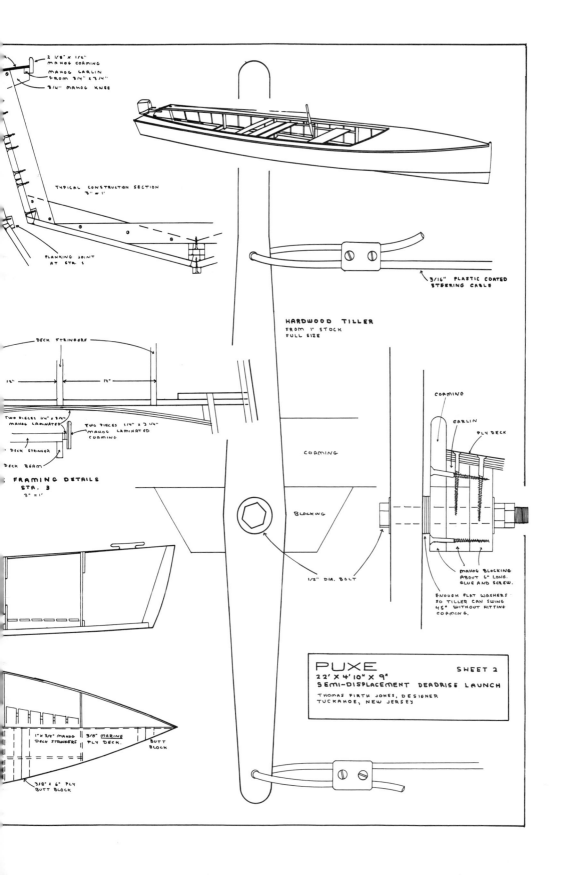

2 1/8" X 1/2"
MAHOG COAMING

MAHOG CARLIN
FROM 3/4" X 3/4"

3/4" MAHOG KNEE

TYPICAL CONSTRUCTION SECTION
3" = 1'

PLANKING JOINT
AT STA. 5

DECK STRINGERS

1"   12"

TWO PIECES 1/4" X 3/4"
MAHOG LAMINATED

TWO PIECES 1/4" X 3 1/4"
MAHOG LAMINATED
COAMING

DECK STRINGER

DECK BEAM

FRAMING DETAILS
STA. 3
3" = 1'

HARDWOOD TILLER
FROM 1" STOCK
FULL SIZE

3/16" PLASTIC COATED
STEERING CABLE

COAMING

BLOCKING

1/2" DIA. BOLT

COAMING

CARLIN

PLY DECK

MAHOG BLOCKING
ABOUT 6" LONG.
GLUE AND SCREW.

ENOUGH FLAT WASHERS
SO TILLER CAN SWING
45° WITHOUT HITTING
COAMING.

1" X 3/4" MAHOG
DECK STRINGERS

3/8" MARINE
PLY DECK

BUTT
BLOCK

3/8" X 6" PLY
BUTT BLOCK

PUXE   SHEET 2
22' X 4'10" X 9"
SEMI-DISPLACEMENT DEADRISE LAUNCH

THOMAS FIRTH JONES, DESIGNER
TUCKAHOE, NEW JERSEY

creeks, all sooner or later feeding into Great Egg Bay. Here distances are great because the rivers are serpentine, and the marshy scenery is pleasant but not varied. Tidal currents can be as much as 2 knots. In a boat 20 feet on the water, hull speed is 6 knots, which meant the scenery would sometimes be going past at 4 knots. We doubted we'd be satisfied with that.

Hull Speed

The reason a boat has a hull speed is that it pushes aside water. The water at bow and stern is forced into motion at the same speed the boat is traveling. The laws of physics dictate the relationship between the speed of water in a wave and the distance from one wave crest to the next. Hull speed is equal to the square root of distance (waterline length) times 1.34.

There are three ways to exceed hull speed: One is by surfing or coming down the face of a wave that you have not generated yourself, using power supplied by gravity. Usually surfing is of short duration, but a Portuguese friend of ours once set out on his surf board to follow the inter-island ferry *Espalamaca*, famous for her giant wake. He quit at the Horta harbor mouth, but was sure he could have made it all the way to Pico, if he'd wanted to go.

The second means of exceeding hull speed is by planing, or applying so much power that the boat is lifted up over its own bow wave. Then, like an airplane, you are using much of your power to stay aloft, and only the remainder to move forward.

Ten years ago, Bill Durham wrote an article in *Small Boat Journal* about what he called "auto boats," both because they use converted automobile engines and because they have "the same size and shape as a car, and the same power density, though they won't go so fast." However, Durham never explained the great attraction of these boats, which is that in going from his car to his boat, the owner need not make any change in his thinking. Ashore, he gets into his car and drives to the drug store, one mile away, at 50 mph. Afloat, he drives out to the fishing ground, one mile away, at the same speed. The arrangement of controls in his boat, the whole feel, is as much like his car as possible.

If you anchor near the fishing ground, you may exchange nods with a calm, quiet fisherman. Perhaps he is tooling around under electric outboard, looking for the perfect spot. Suddenly he's tired of it, or he remembers that it's dinnertime. Down comes the outdrive and away he goes like a scalded cat, throwing a wake four feet high, toward a dock a few hundred yards distant. It is so bizarre that no matter how often I see it done, I'm always surprised. The nice fisherman has never seen it done—or thought of the possibility of doing it—any other way.

Durham is a steam-power buff, but he will settle for internal combustion engines if they turn slow enough. He points out that in the early days of motorboating, skippers were interested in their engines and propellers, and saw as much beauty in them as in their hulls. But he's also honest enough to admit that the low compression ratios of slow-turning engines cost them all the efficiency that they gained with slow-turning props. He also admits in a *Mariner's Catalog* that steam engines are about a third as fuel-efficient as internal combustion ones. Though I have occasional hankerings for slow-turning engines, which are such fun to look at and listen to, one has to be more of a buff than I am to put up with their idiosyncrasies.

In the beginning, there was no other kind of engine. Boats couldn't plane because one horsepower is needed to lift every 50 pounds, and the engines alone had power-to-weight ratios hardly better than that. Designers had no choice but to use the third and most effective way of exceeding hull speed: They made their hulls so narrow and sleek that they *cut through* their own bow waves, rather than ride up and over.

There are limits to the speed available by cutting through your own wake. Multihull sailboats do it, and until recently a multihull held the world record for speed under sail. Now the record is held by a windsurfer which is a planing hull, and I doubt a multihull will ever hold it again. But for transporting a reasonable load on the water in a variety of conditions, no planing sailboat will ever touch a multihull.

Weston Farmer devotes many pages of his wonderful book, *From My Old Boatshop*, to long, narrow powerboats. About twice hull speed, he thinks, is a reasonable expectation if waterline length-to-beam is at least 5 to 1. He wants the bow half-angle fine but not pinched (no hollow in the load waterline), and he says one horsepower will drive several hundred pounds of boat—say 10 horsepower per ton. This is still overpowered by Bill Durham's standard. Two to five horsepower per ton is normal power for a pleasure boat, Durham says, and boats so powered have the look and feel of a boat. Give them 100 horses per ton, and boats inevitably have the look and feel of automobiles.

Boats for special purposes (e.g., search and rescue boats) can sometimes make use of the power of an auto boat. Here, the crew is being paid to put up with the nasty look and feel of the boat. Recreational powerboats seldom have such requirements. The biggest problem with a recreational powerboat is finding anything to do with it. Usually the owner wants to spend a certain amount of time on the water, and if he isn't a fisherman, he has no idea where to go or what to do. The solution, obviously, is not to do nothing fast, but rather to do nothing pleasantly. And to do it in a boat that feels like a boat.

Puxe was designed to make best use of the 7.5- or 10-horsepower Honda outboard, which was at that time the only 4-stroke outboard on the market. Farmer's long, narrow powerboats all have inboards. For shallow water that is often full of grass, however, an outboard is my choice. It costs more than an inboard, but not when you consider that you're getting drive train and steering as well as engine. Four-stroke engines are very much quieter than two-strokes, and that's an excellent start on pleasure boating. They also get almost twice the gas mileage, and even if you can afford to burn gas and have no environmental qualms about doing so, you may not enjoy carrying it to the boat.

In long, narrow powerboats, Farmer prefers a round bottom. Nevertheless, a semi-displacement hull, like a planing hull, is poorly supported aft once it moves out through its bow wave and ahead of its stern wave. It needs a broad lifting surface aft, so the bilges must be tight bends, which are hard to make in many kinds of construction, and are often not durable. Chined hulls are simpler to build and stronger. They give more stability to a narrow boat. At speeds above hull speed, chines don't increase resistance.

Tow-Testing Models

When first thinking through this design, I wondered if a flat bottom wouldn't suffice. It would be lighter and easier to build than a warped-V. Deciding temporarily on a boat 20 feet by 4 feet on the water, I made models at 3/4-inch equals one foot, or 1/16 scale. They were ballasted to weigh 1250 pounds divided by 16 cubed, or 5 ounces. Test speed of the models would be the square root of 1/16 times the speed of the full-size boats, so hull speed would be 1 1/2 knots and desired cruising speed 2 to 3 knots.

Using a light stick with brads driven through it at measured points as a balancing arm, I tested the models against each other in the river current. At less than hull speed, the flat bottom may have had slightly less resistance. At 2 knots (the fastest current I could find and reliably measure), the flat bottom had nearly twice the resistance of the V-bottom. In fact, the flat bottom would not exceed hull speed at all without lifting its forefoot and trying to plane.

The long, narrow, warped-V hull is the traditional workboat of the Chesapeake, where it is locally known as a "deadrise" hull. With the introduction of cheap power from wrecked automobiles, some of them grew fatter to get more stability and loading capacity at the cost of efficient movement through the water. But if you go today to the state marina at Kent Narrows, Maryland, you'll see a hundred deadrises—some wood and some fiberglass. They are used for commercial crabbing and oystering, and most still have the narrow proportions of low resistance.

Puxe is 22 feet by 5 feet overall, 21 feet by 4 feet on the water. She draws 9½ inches and displaces 1250 pounds. The hull weighs 450 pounds. She follows the well-known rules for chined designs: The less rocker the better, with perfectly straight lines in the aft third of the hull. The transom is ⅞ of the maximum chine beam, and the sheerline in plan is pulled in to give a minimum of flare to the topsides aft. This avoids the nuisance of tumblehome, but still gives some shape to the stern. The deck coaming curves in more dramatically to increase the boaty feel, as well as to stiffen the transom.

The old-fashioned preoccupation with the lines of a boat reaches its apotheosis in the profile-view sheerline. We are told that this line must be the essence of beauty, and thus it will determine the whole character of the boat. The truth is that it is never seen, except in the plans. You never look at a boat exactly from deck level and amidships. Even if you did, perspective would distort the sheerline.

The purpose of the sheerline is to connect the freeboard needed at the various stations. It will be harder to build and look at if it isn't fair, and in a hollow sheer, freeboard requirements will usually dictate more curve aft than forward. I draw all my hollow sheers with a plastic ship's curve (K&E 57 1685 35). Handsome is as handsome does.

With four people amidships, the gas tank as far forward as possible, and a certain amount of lifejackets, paddles, anchor, and other junk under the foredeck, *Puxe* trims well. More than one viewer has remarked wistfully on the "wasted space" behind the helmsman's thwart, but to keep the waterline somewhere near the water, engine *thrust* as well as engine weight must be considered, and accommodation comes as far aft as I dared put it. We don't mind stepping over the low thwartback, which is fixed and probably needs to be to stiffen the long, open cockpit. The forward thwart lifts out, and the sole is long enough for camper-cruising, though we've never tried it.

Steering is by whipstaff, which is traditional to deadrises and is comfortable to use whether the helmsman is standing or sitting. For a vanity, the staff is varnished walnut. With one person aboard, she can be steered by walking from side to side in the bottom, and it's fun to do—something like steering a sailing kayak without using a paddle. It has always seemed less trouble, as well as cheaper, to go aft to shift gears or adjust the throttle, rather than to buy remote controls and find a way to install them. Early on, I realized it would be safer to run wire for the kill button up to the steering station. Eleven years later, I'm still planning to do it. In tight spots, we do go aft to steer.

Puxe had her first launching day in April, with the trees still bare and the weather cold; but the way this narrow boat cut through the water warmed us all. Our hand-held speedometer only read to 7 knots, but we pegged it with a very small throttle opening. Route 9 crosses Great Egg Bay not quite 10 nautical miles from our dock, and at half throttle we are under the bridge in an average time of 55 minutes, though it does vary with the tidal current. At half throttle, the Honda burns .4 gallons per hour.

Top speed may be 12 knots in flat water, but it's noisier and we seldom do it. Full throttle is drier to windward in a chop, and 14 knots may be available then, with two crew sitting as far aft as possible. A good deal of the keel is out of water, and the boat is operating as a semi-planing hull, which is a different type altogether. The semi-planing boat has a planing shape without quite enough power to level out and plane, and this is the usual performance of powerboats with accommodation. *Puxe* can get 14 knots in a chop, because both the angle of trim and the broken water passing under the boat reduce wetted surface, which is about 25 percent of total resistance at this speed. The Honda is burning about a gallon an hour, so miles per gallon are about half what they are at cruising speed.

Puxe has no skeg, but she runs straight in protected water. In turns, she is obedient and mannerly. With two people aboard, she leans into the turns; with six, she leans out (and speed is down to 8 knots). She is too long to be an ideal boat for narrow creeks. She can handle wakes from other boats—most comfortably on the beam—and the wind and chop of bays several miles wide on days when small craft advisories are posted. Like Ralph Clayton's Adams garvey, she is not a good boat for the ocean. Downwind in even moderate seas, the stern tries to pass the bow, first on one side and then the other. The motor revs up and down. It's the same in any powerboat with a flat or submerged transom: In a wave train and going slower than the waves, they are not fun, and perhaps not safe. Going faster than the wave train would bring a different but no less serious set of problems. Semi-displacement boats are no more safe or pleasant to use in the ocean than semi-planing or planing forms. In calm weather, any of them will do. In rough weather, none is worth much.

Puxe was designed for sheltered water, and few of my boats have done their intended job as well as she has. She throws so little wake at cruising speed that we are never embarrassed to use her near docks or other boats. She is endlessly pleasant to be in, especially now that she has a Bimini top, and we look forward to those stinking hot days when work is impossible and we have an excuse to spend the day in her. She gets under way very quickly, and unlike a sail or paddle boat, she can be taken out for very short cruises—for half an hour before dinner—with perfect confidence that she will have us home by the time the chicken is ready to come out of the oven. She is so cheap to run that we never hesitate to start her up. We have kept her longer than any other boat.

She is almost as small as a 5-to-1 boat can be and still be stable enough to use carelessly. You can stand or walk around in her anywhere, provided not too many people do it at once. But I wouldn't want an open boat much longer. It would take another 4½ feet of waterline to get another knot of cruising speed at the same speed-length ratio, and the extra weight and wetted surface would require a bigger engine.

She does have three or four minor shortcomings. One is her 100 pounds of soakage, which makes her hull weigh 550 pounds each fall. This increases displacement 8 percent, which is not crippling to performance and does not make her too difficult to pull out on rollers in November. It does make her heavy to turn over for the winter. We could, of course, block her up, cover her, and leave her right-side up; but wooden boats like to have good air circulation, even in the cool months, and covers usually sag or leak, as well as block the air.

At first, with four people aboard, she put up a constant stream of spray from the bowpost, a piece of guard metal 3/4-inch wide. Lighter loads keep the bowpost just out of the water, and heavier ones reduce speed enough to keep spray down. I made a funny-looking whisker, just above waterline and coming around the bow and aft 6 inches. This knocks down the spray, but I don't like these add-ons, and would rather have a hull shape that didn't make spray in the first place.

With only two aboard, *Puxe* doesn't have her whole length in the water. She lifts her forefoot an inch or two above the surface, which costs her a foot of waterline. She runs just fine, but I keep wondering if she is operating in a true semi-displacement manner when lightly loaded. This could be corrected by moving the gravity forward, or the buoyancy aft.

Puxe II

This boat was intended to remedy *Puxe's* few shortcomings, and to be yachtier in form and finish, but little different in size or purpose. The cuddy is only big enough for a portable toilet, and even then, the user had better be small or supple. The chines on *Puxe* rocker up to meet the waterline at the stern, but in *Puxe II* they are horizontal, giving more buoyancy aft. The sections forward of midships are hollow, as they often were on motor launches in the 1920s, giving a finer entry and less buoyancy forward. The center of buoyancy is 13 inches further aft. However, the reserve buoyancy forward is greater than in *Puxe*, and this and the clinker planking should make her a drier boat. The forefoot is rounded in profile,

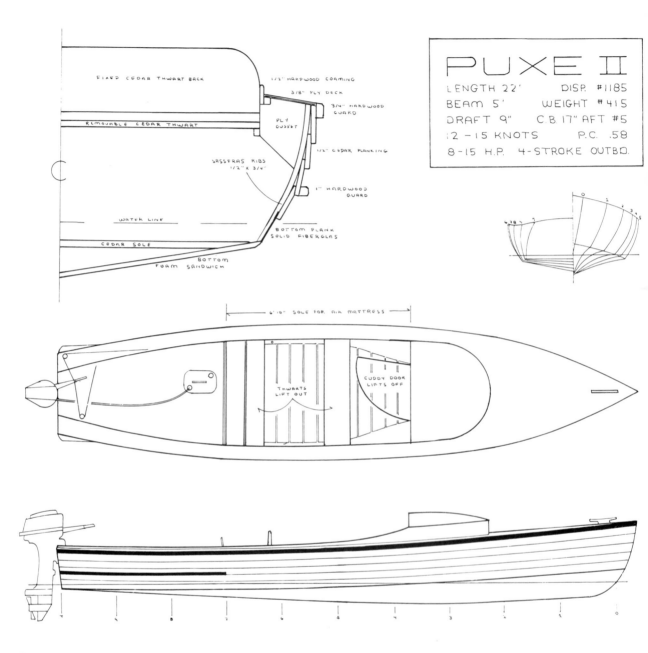

FIXED CEDAR THWART BACK

1/2" HARDWOOD COAMING

3/8" PLY DECK

3/4" HARDWOOD GUARD

REMOVABLE CEDAR THWART

PLY GUSSET

1/2" CEDAR PLANKING

SASSFRAS RIBS 1/2" X 3/4"

1" HARDWOOD GUARD

WATER LINE

CEDAR SOLE

BOTTOM PLANK SOLID FIBERGLAS

BOTTOM FOAM SANDWICH

PUXE II
LENGTH 22'   DISP. #1185
BEAM 5'   WEIGHT #415
DRAFT 9"   C.B. 17" AFT #5
12-15 KNOTS   P.C. .58
8-15 H.P. 4-STROKE OUTBD.

6'10" SOLE FOR AIR MATTRESS

THWARTS LIFT OUT

CUDDY DOOR LIFTS OFF

Puxe II.

and in no state of loading should she throw a stream of spray up toward the foredeck.

Her bottom is foam sandwich, and at the chine, her outer and inner skins are joined to make the first plank of the clinker topsides. The upper four planks are 1/2-inch cedar. Clinker can be lighter than carvel, and *Puxe*'s planking is 3/4-inch. The clinker planks were hung over sawn frames on about 3-foot centers, which were bolted through the lowest, fiberglass plank. Between the sawn frames are bent frames of steamed sassafras. Fastenings are bronze screws and polyurethane caulking. Decks are 3/8-inch mahogany marine ply.

The round front of the cuddy was vertically strip-planked in two halves in a female mold. This went quickly. I started to cold mold the cuddy top, but found the work so aggravating that after the first layer of veneers was on, I glassed it lightly to solidify it, and used it as a male mold to make a solid glass cuddy top. Cellophane is a good parting agent for a male mold like this, because it stretches and lies down flat.

The tumblehome aft is for looks, of course, and it does look pretty, but it requires double guard rails. The clinker planking is also for looks, though it is lighter than carvel, and may knock down some spray. The big weight saving is in the bottom. I believe *Puxe II* is 75 pounds lighter than *Puxe* in the spring, and 175 pounds lighter in the fall. Fiberglass does absorb water, enough to blister the gelcoat in some production boats, though nobody seems to agree on how this happens.

The best current answer to osmosis in fiberglass boats seems to be epoxy coatings, though copper is sometimes used too. Epoxy resin is more nearly waterproof than polyester, and the marine paint companies are falling over each other to market epoxy bottom coatings at the same bargain prices as the rest of their products. It may be worth using these coatings, rather than epoxy resin, if they flow and spread like paint. Currently, I am using industrial epoxy primer paint on the outsides of fiberglass hulls.

Putting a couple of coats of epoxy paint on the outside of a one-off hull is not much trouble or expense, but for the mass-producer of fiberglass hulls it is a ruinous procedure. The reason most hulls are fiberglass today is not that fiberglass is quicker or cheaper or better than other materials. Rather, fiberglass allows the transfer of finishes without the labor of making them. A builder will spend thousands of hours finishing a plug until it is as perfect as a new refrigerator or automobile, because that labor can be amortized in successive hulls pulled from successive molds taken from the perfect plug. Of course, the finish of the hulls is not durable, any more than hair dye or the other surface finishes that most people want these days, but it does sell boats. Finish is the modern substitute for

Figure 4–6. The original customs launch in Bridgetown, Barbados.

workmanship, and the production builder cannot afford to ruin his
molded finish with a coating that must then be finished by hand.

Puxe II looks to me like an improvement on old *Puxe,* especially if you
like your boats on the fancy side. I have built one example, but have not
yet had a chance to try her out.

Barbados Customs Launch

Shortly after Carol and I anchored in Bridgetown, Barbados, following our
first tradewind crossing in 1979, we saw a very long, narrow boat weaving
through the fleet toward us. The cockpit was forward and the cabin aft, in
a style common in British river launches 60 years ago. The cabin top was
highly cambered. The side of the boat was hung with tires. Despite the
heat, the pilot and his mate wore dress shirts and long, black trousers.

With the big, slow-turning engine barely ticking over, the launch eased
alongside us. The two men were smiling, relaxed, and friendly, and obviously very pleased with their boat. In soft voices and with great courtesy,
they said that they were from the customs service and had come to clear
us. Our last experience with customs had been in Spain, where everything is done with a maximum of noise and hurry and a minimum of
courtesy. It would be hard to say which pleased us more, these two gen-
tlemen or their vessel.

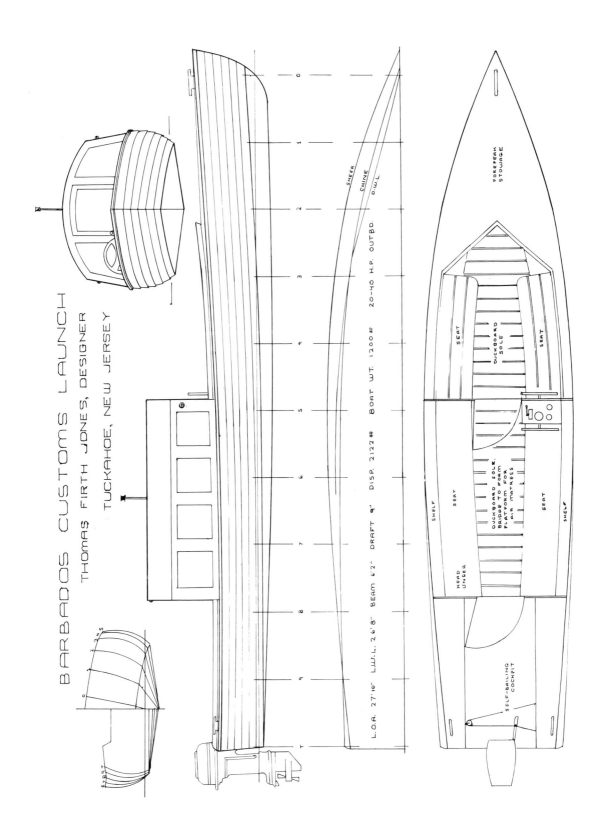

BARBADOS CUSTOMS LAUNCH

THOMAS FIRTH JONES, DESIGNER

TUCKAHOE, NEW JERSEY

L.O.A. 27'10" L.W.L. 26'8" BEAM 6'2" DRAFT 9" DISP. 2122# BOAT WT. 1200# 20-40 H.P. OUTBD.

SHEER

CHINE

D.W.L.

FOREPEAK STOWAGE

SEAT

DUCKBOARD SOLE

SEAT

SHELF

SEAT

DUCKBOARD SOLE, BRIDGE TO FORM PLATFORM FOR AIR MATRESS

SEAT

SHELF

HEAD UNDER

SELF-BAILING COCKPIT

The boat had a round bottom and a true semi-displacement form. She was painted in work-boat colors, but she still seemed yachty and may well have started life as a pleasure craft. On the Thames, where this kind of layout originated and where the weather is almost always bad, the owner and his guests customarily sat in the cabin, while the paid hand ran the boat from the cockpit forward. Some old limousines followed the same principle: The owner better appreciates the luxury of the back seat if the chauffeur is out in the snow. But in Barbados, the two customs men were on deck as much as possible, and ducked below only to grab another sheaf of papers.

The cockpit was largely occupied by the engine box, and the passageway between it and the coaming was probably too narrow for two men to pass each other. Controls were forward, and the helmsman was close to the noise and heat of the engine, with no provision for comfortable seating. The cabin was too short to be a very useful space, though it did have standing headroom on centerline. For some reason, the aft deck was flush—perhaps big tanks were located under it. For her length, the launch had appallingly little useful space in her, but that didn't detract from her grandeur, and may even have added to it. We looked her over as much as we could, and took some pictures.

Puxe was under construction a year later. She was derived from Chesapeake workboats rather than the Barbados Customs Launch, but Carol and I often remembered that boat, and sometimes we joked about putting a tiny cambered house on *Puxe*, two-thirds of the way aft. Six years later, we were still talking about the launch from time to time, and I decided to build a boat that had her layout and looks, but with more usable space and the proven hull shape and building methods of *Puxe*.

She was drawn twice the size of *Puxe:* Linear dimensions were therefore multiplied by the cube root of two. This yielded a displacement of 2500 pounds, which seemed too much, because the payload in a boat seldom needs to go up in direct proportion to the size. *Puxe* has an 800-pound payload, and it seemed unlikely this boat would need more than 1000 pounds. Twice *Puxe's* hull weight made 900 pounds, and the cabin and furniture might add another 200 pounds. To bring the displacement down nearer to 2100 pounds, the deadrise of the bottom was made shallower, so that the 28-footer draws no more than the 22-footer. Overall beam is 6 feet 2 inches. Waterline length and beam are 26 feet 8 inches by 5 feet 6 inches, making her just shy of a 5-to-1 boat. This compromise was needed to stabilize the higher center of gravity that comes with a cabin and a self-draining deck aft.

At 10 knots, *Puxe* runs at a speed-length ratio of 2.2. At the same ratio, the Barbados Launch would cruise at 11.3 knots, but would need almost twice the horsepower to do it. There was no 20-horsepower, 4-stroke outboard on the American market, so the choice would be between a 20-horsepower 2-stroke, which would burn nearly four times the gas of the Honda in *Puxe*, or the 15-horse Honda that had just come on the market. In a boat proportionately lighter and shallower, it might just be enough. At any rate, it would be the owner's decision, and noise would be little problem, with a cabin or at least a door always between crew and engine.

As was later done in *Puxe II*, the forefoot was rounded and the chines aft were horizontal. However, the boat was built entirely of wood. Even in this size, half-inch clinker planking is enough for topsides, and the bottom is 1-inch square cross strips, which went on gratifyingly faster than ³/₄-inch strips. I doubt it took much longer to lay the bottom on the Barbados Launch than on *Puxe*. The cabin top was also strip planked, and the cabin sides and decks were plywood. In a boat where appearance is as important as in this one, there is tumblehome, of course, and there are the double guards that go with it. The cockpit coaming is one of the few concessions to practicality. These V-shaped, forward-sloping coamings are less handsome than round coamings, but far drier.

I thought the layout pretty clever. Eight people could sit in the cockpit if need be (the Coast Guard, when licensing passengers-for-hire boats, figures the average bum at 18 inches), although the boat would clearly be

Figure 4–7. Barbados Launch in frame.

Figure 4–8. Barbados Launch nearing completion. Lofting still on wall.

below her marks at those times. The cabin had only 4½-feet of sitting headroom, but was nearly 8 feet long, and I figured a very comfortable double could be made by bringing up the sole to ledgers on the seat fronts and rolling out a mattress. A useful deck on the outside of the cabin would have made the accommodation agonizingly narrow, and there had to be a center aisle when the boat was under way because it was the only route from cockpit to engine.

The main control station was meant to be inside the cabin, complete with wheel, gauges, switches, and levers. Steering was to be by cables over a drum, and the drum shaft was to go through the bulkhead to another wheel in the cockpit. From there, the skipper would have to reach in the open doorway to regulate speed, or might have to go in for a moment to read a gauge or flip a switch. Skippering a powerboat is not taxing, and a chance to move around is often welcome. At either steering station, the helmsman would sit facing athwartships, as in a sailboat.

Hand surgeon Louis Benton, who bought the Barbados Custom Launch for a commuter to his summer house in the Thousand Islands, had other ideas. At the helm, he wanted to sit in the cockpit, facing forward at a console. Would I build him four forward-facing seats, instead of the benches, and leave the console to be put in by someone else? Yes, I

Figure 4–9. The oar (right) was the idea of Louis Benton (left). Not my idea of recreation.

would. Would I leave all the furniture out of the cabin, and not paint it inside? And how much would that reduce the price? Louis drove a hard bargain, but has since become a good and generous friend. The act of buying diminishes us all.

Louis came down with a trailer—also a bargain—to pick up the Barbados Launch. Evidently the trailer builder hadn't known how to read blueprints, because the boat would have dropped between the rollers from bow to stern. It was 5 o'clock on a Saturday afternoon, and we had nearly 100 new holes to drill in the steel trailer frame. We all got to it, and Carol and I found it interesting because we had never worked with a surgeon. He put his back into those holes, and drilled them fast, but soon I had to tell him that I could smell his drill, and he'd better ease off. He said okay, but went right on pushing. When the drill stopped turning, he asked for another, and we told him we had only two more left. He looked at us as if we were nurses telling him there were no more sutures. "We'll get some more," he said.

"The stores are closed, Louis," we said. He looked at us with disbelief, and went on drilling. In the end, I spent all my time sharpening bits, hoping to take the load off the drill motors that way. When the last hole was drilled, the last drill motor wouldn't have drilled two more. Louis took the boat away. He paid for the drills without hesitation.

A year later, when we visited the Thousand Islands, Louis and his wife Patricia were extraordinarily kind and hospitable to us. They put us up at their house on the island, introduced us to their friends, and took us to see the antique-boat museum at Clayton. We spent many hours aboard the Barbados Launch. We had no speed measuring equipment, and the current of the St. Lawrence always runs the same direction, making it hard to figure speed against buoys or points ashore. As soon as we were away from the dock, Louis opened the throttle right up on the 20-horse-power Evinrude, and left it there. Despite that, we didn't think she went any faster than *Puxe*. She did exceed hull speed (about 7 knots in this boat) and did not appreciably change trim to do so. A big boat always goes along with less fuss than a smaller one, and possibly we were fooled; but we thought 10 knots was about all she did.

I've pondered this often, and I can't imagine that changes in hull shape are responsible. They are so minor: length-to-beam is 8-percent less fine, center of buoyancy is 4-percent further aft. Bow half-angle is the same at 14 degrees, and the slightly immersed transom of the Barbados Launch makes the P.C. of this boat .55, while *Puxe* has only .53. Both of these are generally considered too low for the speeds the boats go.

Louis has not overloaded his boat, especially as I think she weighed less than 1000 pounds when he took her away. He has a Bimini, and so would I, though it doesn't improve the looks. He has several fenders which are often dangling in the water, but they don't make a big wave train. He does have a heavy battery aft for starting the engine, and he keeps the gas tanks closer to the engine and transom than I would. But the farther-aft center of buoyancy should compensate for that. He has a minimum of gear under the foredeck, and the cabin is still bare of furniture.

I think the propeller is the problem—not the pitch, but the diameter. Around the boatyard, you're likely to hear the old salt say that one inch of diameter is equal to two inches of pitch. This is a near-truth. One inch of diameter may absorb the energy of two inches of pitch, but diameter is diameter and pitch is pitch. There is nothing equating them, or making one do the work of the other.

Weston Farmer has a rule of thumb, inherited from the steamboat days: "A propeller, to command a boat, should have a disk area equal to one-fourth the area of the immersed midsection." The Honda 7.5 in *Puxe* has a 9½-inch prop, which is 30 percent of the hull's midships section. Louis bought his Evinrude because it was a leftover, and cheap. It has a 10-inch prop, 23 percent of midships section. Though either might do for a displacement steamboat, I suspect that for a semi-displacement hull, both are too small and the Evinrude prop is *much* too small. This can't be changed much, because the engine leg doesn't have room for a much bigger prop. This is one of the disadvantages of outboards.

Both engines peak at 5000 rpm, but the gear reduction is greater in the Honda, which is one reason it can swing a proportionately larger prop. According to the nomograph in *Skene's*, the Honda prop is half an inch too big for its horsepower and rpm, and the Evinrude prop is only an inch too small. But *Skene's* goes on to say that prop size must also be scaled to boat speed and displacement. The Evinrude prop is meant to push a light, open boat (the smallest of auto boats, in fact). If *Puxe* does 12 knots with throttle wide open, the speed-length ratio is 2.6. At the same ratio, the Barbados Launch should do 13½ knots. If she only does 10, that's mighty disappointing.

Whatever speed the Evinrude pushes her through the water, the Barbados Customs Launch is a real looker, and that was the idea in the first place.

Long, Narrow Powerboats

5. Sailing Pocket Cruisers

Of all the boats in this book, the ones in this chapter least deserve to be called low resistance. *Mock Turtle*, designed by K.V. Weisbrod, is the best of them and *June Bug*, designed by William Garden, is the worst. My two designs fall somewhere in between. To my mind, nothing less than sitting headroom constitutes accommodation, and if a pocket sailboat is to have both accommodation and low resistance, it needs to be a pretty long pocket. All these boats are too short and bulky for low resistance, but some have charm, and some are fun to sail just because they are such short cruisers. During the hundred months I've had my shop, it's been occupied by these boats for 15 or 20 of them. I'd like to show them to you.

Low resistance is important to all boats because, lacking it, the only way to get acceptable speed is with more power. Low resistance is the product of good design and, to a lesser degree, good construction. Low resistance often costs less than high resistance, because power is expensive, and will become more expensive in the years to come. Maybe some day, even the deep-V crowd will have a bellyful of it.

Low resistance is especially important to a sailboat. Even if the power can be afforded, it often cannot be applied, because the sailhandling platform isn't big enough to work it or because the hull isn't stiff enough to stand up to it. Bill Durham says, "One thousand square feet of modern sails will reliably deliver from one horsepower (running in a 10-knot breeze) to 9 horsepower (reaching in a 20-knot wind)." But it's hard to think of putting 1000 square feet of modern, controlled sail on a hull less than 45 feet overall, and if she has a long keel or a fat midsection with low prismatic, she still may not go.

The throne of "Marine Schmaltz" has many claimants: William Atkin and Jay Benford are two who immediately come to my mind. But of them all, William Garden must be the true king. Garden can write reasonably lean, though his conclusions don't always follow his facts. The boats he has designed for himself are almost all commendably lean and purposeful. But when it comes to designing for others, Garden never hesitates to put (as he only half-jokingly says) a "plastic lobster in a wicker basket." This has made him the favorite designer of those Taiwanese builders who bed nothing, use solid brass hardware, and drive screws with hammers. I once worked on such a boat, not two years old, where you could put your hand through the cabin side almost anywhere. Never mind, while they lasted, they looked like pirate ships indeed.

June Bug

The Rudder commissioned this design from Garden just as the second World War was ending. *Rudder's* policy was to pay the designer a one-time fee, print the complete plans to a reduced scale in an issue of the magazine, and sell the full-size plans for the cost of blueprinting. "For the benefit of the sport of yachting," it did not attempt to place restrictions on their use. In addition to Garden, designers included Crocker, Crosby, Fox, L. F. Herreshoff, Mason, Mower, Steward, and Wittholz. I bought a bunch of these plans before *Rudder* went belly-up, and wish I'd bought them all.

Garden was a young man in those days, but his manner was already well developed. He gave the editors some salty talk about "the young fellow who likes to cruise." The construction of *June Bug*, he said, "isn't beyond the capabilities of a young fellow handy with tools." *Rudder*

Figure 5–1. Bobby and Gail on June Bug *in Great Egg Bay.*

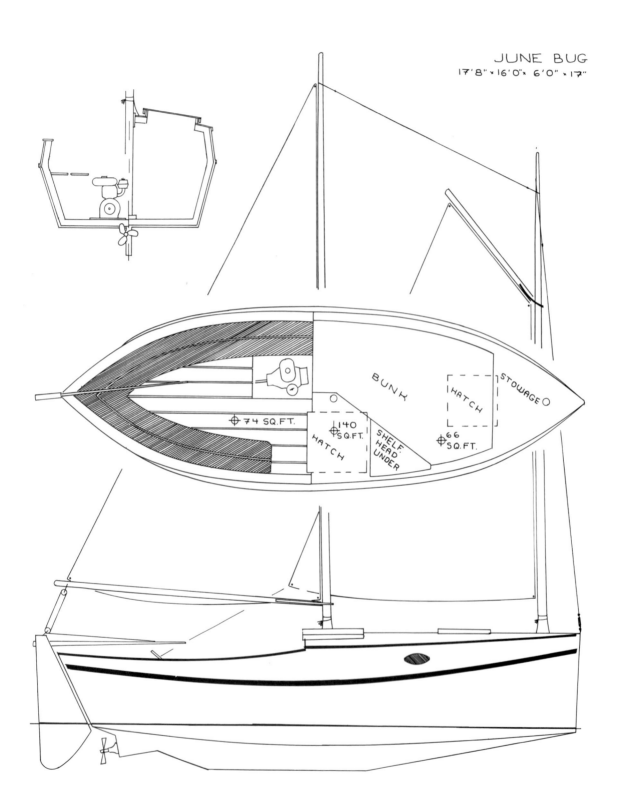

JUNE BUG
17'8" × 16'0" × 6'0" × 17"

74 SQ.FT.

140 SQ.FT.

HATCH

SHELF.
HEAD
UNDER

BUNK

HATCH

STOWAGE

66 SQ.FT.

advertised the plans as "designed for boys having limited funds and tools. She affords rugged cruising accommodations for two adventurous lads." Anyone who has tried building this boat has discovered that unsteamed wood doesn't particularly like some of the curves that Garden has drawn, especially in the topsides at the stern, where he is trying to get some bearing to carry sail. All in all, the project is nowhere near as easy as it looks.

June Bug is a dory-hulled cat schooner, 17 feet 8 inches by 6 feet on deck, and 16 feet 4 inches by 5 feet on the water. There is no keelson. The shallow external keel of $1^5/8$-inch oak runs the length of the bottom, and provides both leeway prevention and structure. She has seven sawn oak frames with cleats between them. She is carvel planked, but Garden doesn't say with what, and the scantling rules in *Skene's* weren't written for dories. With 12 inches between fastenings, I dared go no lighter than $3/4$-inch cedar.

Traditional dories the length of *June Bug* are built right-side up. The bottom is cleated together and sprung to the desired rocker—perhaps with two horses and a prop to the shop ceiling—and the sides are then put on. These dories don't have keels or chine logs, and their bottoms aren't 5 feet wide. Nevertheless, I decided to build *June Bug* right-side up, because the keel offered such a stiff building platform, and because I'd never done it before. Frames were set up on the keel, and braced to the ceiling. Longitudinals were bent from green oak.

The bottom was some trouble to put on. Even a little deadrise would have allowed a good deal more room to work, but a perfectly flat bottom needs to be at least 30 inches off the floor to swing tools between it and your head. And even then, the sawdust will still be coming down into your face, and the topsides may be so high off the floor that you need staging to hang them. For these reasons alone, I wouldn't build another flat-bottom hull right-side up.

The worst was yet to come. When the lowest topsides planks were hung, I was under the boat fairing them off with a Porter-Cable electric block plane. I am left-handed, and the box this tool came in says that it can't be used in the left hand. Lying on my back too close to the work and with a face full of wood chips, I reached over with my right hand to turn off the plane, and put two knuckles of one finger into the blades. Among life-long woodworkers, probably half have lost a finger or a part of one; but I have found this loss damaging in more ways than just physically. Don't do it.

June Bug's cabin is small, and Garden filled the bow of it with a cast-iron stove, leaving 6 feet of sole for sleeping bags to be rolled out. This might do for "young fellows," who might even stomach the heat, smoke, and dirt of a coal stove; but my imagined customers would more likely be a couple, who would prefer to sleep in a double bunk, dine ashore, and stay home in cold weather. I gave her a trapezoidal but serviceable bunk, big enough for adults, with a small flat beside it and a bucket head underneath.

I made almost all *June Bug's* hardware myself. I was still working for another builder then, though I could see his economic curtain descending, and the *June Bug* was a way of getting started on my own. There was never a time sheet on the wall. That's the way to build anything, and it's the way to live, too. Time and money are not the same thing, and what makes work worth doing is not the profit or even the product, but the personal growth and satisfaction. For this boat, I made not only cleats, but blocks and other fittings, too. A friend with a newly acquired lathe turned out the belaying pins for the foresail sheets, though he had to be restrained from adding extra grooves and flourishes. A lathe is so much fun to use that the most usual products are toothpicks and sawdust.

Inboard Engine

Garden called for a 2-horsepower Palmer marine inboard engine. Except for the 2-stroke make-and-break Lunenberg, these small marine gasoline engines are all gone. Because the compression ratio is lower, a well-designed one is smoother and quieter than a diesel. It isn't dangerous in the cockpit of an open boat, because the movement of air keeps fumes from collecting. It's inherently cheaper than a diesel to buy and install, and if a 4-stroke was producing 2 horsepower, it might burn 4 ounces more fuel per hour. Never mind, there aren't any.

Although outboards don't mount readily on double-enders, *June Bug* is shallow enough to have taken an outboard in a well anywhere in the cockpit. But the engine leg and the well would have added more resistance to a hull that already had plenty, and part of what you pay for in an outboard (steering, and the tilt-up hardware) would have been wasted. I also thought that the boat would sell easier with an inboard engine.

Honda makes a line of 4-stroke stationary engines, and the 5-horsepower model is available with a 2:1 reduction gear and centrifugal clutch. The whole thing weighs 33 pounds, including the one-gallon gas tank that comes bolted to the cooling shroud. The tank should need filling only every five hours or so. *June Bug* didn't need 5 horsepower, but the clutch and reduction gear were attractive. Engine beds consisted of a piece of plywood screwed to the frames. The only other connection the engine needed was to the prop shaft, and this had to be machined because the engine shaft was metric. Complete with running gear, the engine cost about as much as a 5-horse outboard, and hid under a very small engine box.

The cat schooner rig is what makes *June Bug* so charming. It doesn't seem possible for a schooner to be so small; and, of course, a sprit sail, with or without a jib, would have driven her to more purpose. But the sprit would have been no picnic to handle on that high, teetery deck, and unballasted dories don't sail well, no matter how rigged. Garden was right to concentrate on charm. Talking about another, larger cat schooner that he designed later, he says, "Two masts make a ship and remind us that these are all toy boats and variety is one of their main charms." There is enough truth to this to allow us to overlook the fact that *three square-rigged masts* are actually what make a ship.

Writing of *June Bug* 30 years after he designed her, Garden says, "The rig was always a pleasure to use." I found it so too, especially after we eliminated one of the foresail halyards. I don't know how long the head of a gaff sail has to be before it needs two halyards, but for sure it's more than 52 inches. The other rig problem on *June Bug* was the triatic stay, which required that both masts be set up at once. The best solution we ever found for this was to be sure we had at least two "adventurous lads" on hand before we started rigging.

She was a handy sailer, but not a good one. Bobby Graham, who bought her, had owned a number of motorboats, but this was his first sailboat. He would listen patiently to my explanations of what the various lines did and how they should be adjusted. At last he would burst out, "Cap'n, just tell me which one of them ropes is the accelerator-like!" Shamedly, I had to admit there wasn't one.

Seventeen inches of draft do not materially slow leeway on an 18-foot cabin boat. *June Bug's* waterline length-to-beam ratio of 3.25 wasn't too bad, but her bow half-angle of 30 degrees wasn't too good; and leeway plays havoc with length-to-beam ratio (among other things) because the hull is moving diagonally through the water. She must have had 90 square feet of wetted surface, and this just can't be driven by 140 square feet of sail. Her only ballast was 200 pounds of rock under the bunk, held down with chicken wire and staples. This seemed better than Garden's specification of 200 pounds of concrete poured between the frames. The second owner took the ballast out. He said it made no appreciable difference in performance or stability.

With or without this inside ballast, *June Bug* wouldn't stand up to her modest sail. Her pointed stern and tumblehome kept crew close to the centerline. She heels, and when enough of the rudder is out of the water, she comes up into the wind. All this is happening well below hull speed—I doubt she ever saw hull speed under sail. Garden talks about

Figure 5–2. June Bug *in the morning mist.*

this problem in his book, *Yacht Designs.* In a big open sharpie, he says that you must have a ballasted fin keel or 14-foot hiking boards, and he draws a stability diagram to show the righting arm. With ballast, it's 1.7 feet; without ballast, he doesn't even measure it, but labels it "negligible." This has to be even more true of a boat with a cabin.

Yet *June Bug* can be defended. Before we took her down river, she was anchored off our place for several weeks while Bobby learned to sail her. I'm no aesthete, but mornings when I got up early enough I'd look out the window and see her swinging to her anchor with the river mists rising around her, and she'd stop my heart. I'd wonder—I'm still wondering— just what is the place of schmaltz in yacht design?

She motored very well. Overpowered for a displacement hull (probably 7 horsepower per ton with crew aboard), she reached hull speed easily, and the prop wash over her big rudder made her responsive. Still, the long keel kept her from being capricious. Air-cooling is noisy, of course, but in an auxiliary that may be a tolerable fault. Later, on *Dr. Buzzhead,* Bobby tried to quiet the engine by running the exhaust out the transom; but he discovered that most of the noise was coming from the fins.

Bobby handled her well around docks. With the clutch giving him forward and neutral, he waited to make his moves, and never needed reverse. Coming into a dock under sail, timing is everything, but Bobby taught me that time is nothing in a powerboat. Wait and see what happens, because you can always get moving again.

He was a restaurant chef most of the time he owned *June Bug*, and used her mostly in the middle of the day, between morning prep and evening dinner. For a couple of summers, he took her out most days and enjoyed being on the water. And he loved being seen in her.

Dr. Buzzhead, M.D.

Bobby Graham was a jack-of-all-trades. He happened to be driving pilings near our place when *June Bug* was a-building. Generous, garrulous, maddeningly mendacious, he bought four boats from me the first three years I was in business. And he found me customers for five others. I was "Cap'n Jones" and he was "Cap'n Graham." I was glad for the work, and I'm still grateful.

Bobby named the June Bug *Barbara J.* in honor of his wife. On the delivery trip, Bobby refused to reef, sailed her on her ear, and headed up and rammed the bank. After that, Barbara stayed home. Soon Bobby was sailing with another lady, and when he came looking for a bigger sailboat, he was divorced and remarried. With laudable circumspection, he named the new boat *Island Princess*. However, her spars were Douglas fir, and at the local lumber yard where the select fir is kept, some visionary had scratched into the paint of the shed wall, "Dr. Buzzhead, M.D." Picking over the pile, with the rain dripping on the tin roof and the whine of a forklift in the distance, we had time to ponder this allusive monicker, and came to associate it with the boat. We didn't tell Bobby.

He wanted no "bozo boat." Traveling around the waterfront with him or leafing through the magazines, I learned the characteristics of a bozo boat, and tried to put together the characteristics of a non-bozo boat, or perhaps an anti-bozo boat. I was also thinking about what he liked in the Garden design, and guessing how much money he had, or could borrow. I sketched him a bigger cat schooner, round bilged, and with outside ballast. I sketched him a 28-foot, V-bottomed fantail schooner (thank you, Tom Colvin). He looked at the sketches respectfully and took them away with him, but his eyes never lit up.

Then Bobby came to me with a better proposal than any I'd dared suggest. He had an advertisement for a Cornish Crabber. "Hey, Tom," he said, too excited to use my proper title. "Could you scale down one of these so's I could afford it?"

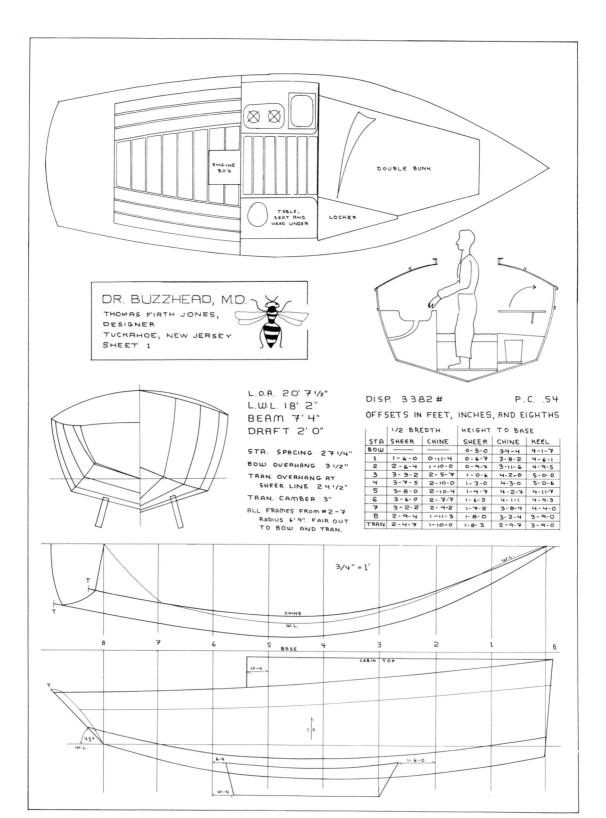

DR. BUZZHEAD, M.D.
THOMAS FIRTH JONES,
DESIGNER
TUCKAHOE, NEW JERSEY
SHEET 1

ENGINE BOX

DOUBLE BUNK

TABLE, SEAT AND HEAD UNDER

LOCKER

L.O.A. 20' 7 1/2"
L.W.L. 18' 2"
BEAM 7' 4"
DRAFT 2' 0"

STA. SPACING 2' 7 1/4"
BOW OVERHANG 3 1/2"
TRAN. OVERHANG AT
    SHEER LINE 2' 4 1/2"
TRAN. CAMBER 3"
ALL FRAMES FROM #2-7
    RADIUS 6' 9". FAIR OUT
    TO BOW AND TRAN.

DISP. 3382 #   P.C. .54

OFFSETS IN FEET, INCHES, AND EIGHTHS

| STA | 1/2 BREDTH | | HEIGHT TO BASE | | |
	SHEER	CHINE	SHEER	CHINE	KEEL
BOW	—	—	0-3-0	3-4-4	4-1-7
1	1-6-0	0-11-4	0-6-7	3-8-2	4-6-1
2	2-6-4	1-10-0	0-9-7	3-11-6	4-9-5
3	3-3-2	2-5-7	1-0-6	4-2-0	5-0-0
4	3-7-5	2-10-0	1-3-0	4-3-0	5-0-6
5	3-8-0	2-10-4	1-4-7	4-2-7	4-11-7
6	3-6-0	2-7-7	1-6-3	4-1-1	4-9-3
7	3-2-2	2-4-2	1-7-2	3-8-4	4-4-0
8	2-9-4	1-11-3	1-8-0	3-2-4	3-9-0
TRAN	2-4-7	1-10-0	1-8-3	2-9-7	3-9-0

3/4" = 1'

W.L.

CHINE

W.L.

8   7   6   5   4   3   2   1   B
BASE

CABIN TOP

10-4

T

C B

45°
W.L.

6-4

1-6-0

10-4

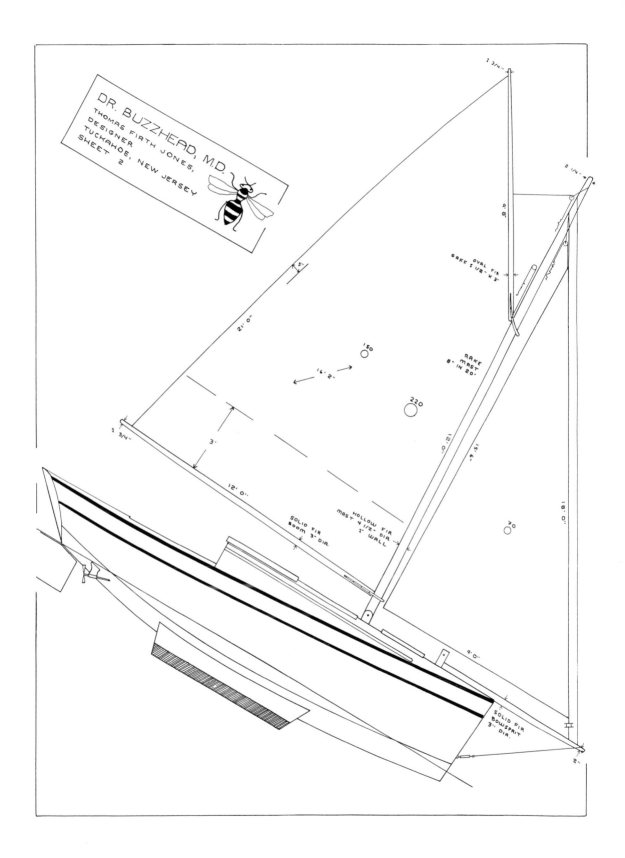

DR. BUZZHEAD, M.D.

THOMAS FIRTH JONES,
DESIGNER
TUCKAHOE, NEW JERSEY
SHEET 2

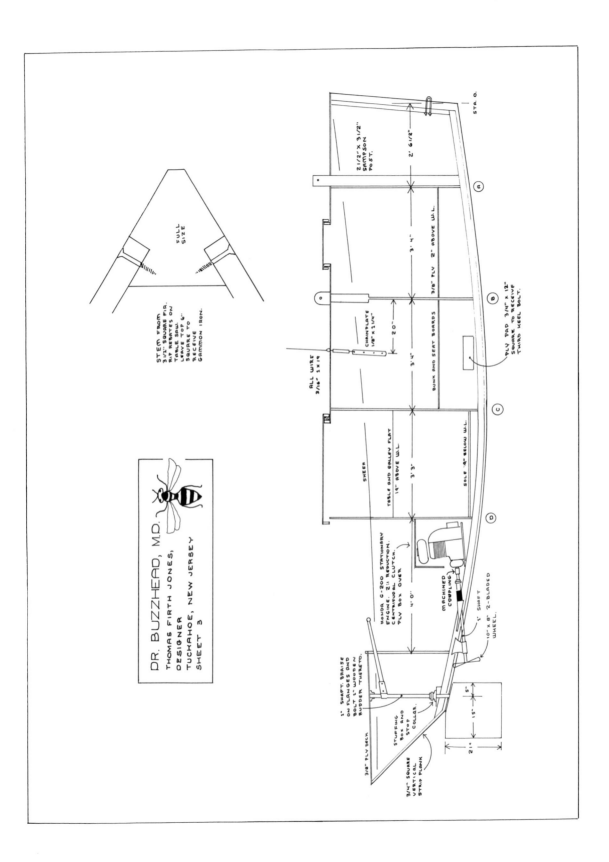

DR. BUZZHEAD, M.D.

THOMAS FIRTH JONES,
DESIGNER
TUCKAHOE, NEW JERSEY
SHEET 3

STEM FROM
3 1/2" SQUARE FIR.
RIP REBATES ON
TABLE SAW.
LEAVE TOP 6"
SQUARE TO
RECEIVE
GAMMON IRON.

FULL
SIZE

2 1/2" X 3 1/2"
SAMPSON
POST.

STR. O.

2' 6 1/2"

A

3' 4"

3/8" PLY 2" ABOVE W.L.

B

ALL WIRE
3/16" 1 X 19

CHAINPLATE
1/8" X 1 1/4"

2' 0"

3' 4"

PLY PAD 3/4" X 12"
SQUARE TO RECEIVE
THIRD KEEL BOLT.

BUNK AND SEAT BOARDS

C

SHEER

TABLE AND GALLEY FLAT
14" ABOVE W.L.

3' 3"

SOLE 4" BELOW W.L.

D

HONDA G-200 STATIONARY
ENGINE, 2:1 REDUCTION
CENTRIFUGAL CLUTCH,
PLY BOX OVER.

4' 0"

MACHINED
COUPLING

1" SHAFT

10" X 8" 2-BLADED
WHEEL.

1" SHAFT, BRAZE
ON FLANGES AND
BOLT 1" WOODEN
RUDDER THERETO.

3/8" PLY DECK

STUFFING
BOX AND
STOP
COLLAR.

3/4" SQUARE
VERTICAL
STRIP PLANK

5"

15"

2 1"

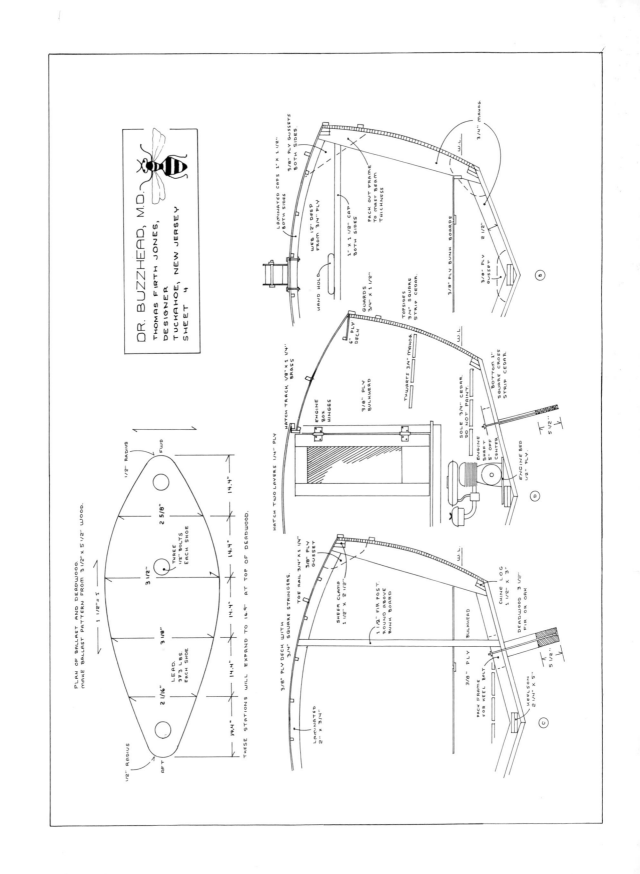

DR. BUZZHEAD, M.D.

THOMAS FIRTH JONES,
DESIGNER
TUCKAHOE, NEW JERSEY
SHEET 4

PLAN OF BALLAST AND DEADWOOD.
MAKE BALLAST PATTERN FROM 3 1/2" X 5 1/2" WOOD.

1 1/2" = 1'

FWD

1/2" RADIUS

2 5/8"

3 1/2"

3 1/8"

2 1/16"

14.4"   14.4"   14.4"   14.4"

THREE 1/2" BOLTS
EACH SHOE

LEAD,
373 LBS.
EACH SHOE

1/2" RADIUS

AFT

THESE STATIONS WILL EXPAND TO 16.8" AT TOP OF DEADWOOD.

Figure 5–3. Dr. Buzzhead, M.D. *sailing in Great Egg Bay. She needn't be reefed.*

The Cornish Crabber is a production English cutter, originally designed for multichine plywood. The builders experimented with chine
lines in the earlier boats, and when they had built a hull that pleased
them, they froze it by popping a mold. Succeeding boats have had a
molded fiberglass hull with a surface that is a succession of flats. This is
not a good idea. In the first place, laying up a chined fiberglass hull is

trickier than laying up a round one. Second, fiberglass is strong but not stiff, and curves help to stiffen it. And furthermore, chined hulls inevitably have more wetted surface than round ones.

The appearance of the Crabber—and appearance was everything to Bobby—was quaint and attractive. Twenty-four by 8 feet and 4600 pounds heavy, she is a flush decker with a gaff topsail cutter rig—plenty of strings to pull. Finished yachty, with centerboard and diesel engine, she cost more than twice what Bobby said he could spend.

English Workboats

In the last days of working sail, most workboats in the southwest of England were sloop- or cutter-rigged. They had nearly plumb bows, with

Figure 5–4. Bow detail. Bobby wanted the chain bobstay. Note copper vent behind Samson post.

long, reefing bowsprits; curved and heavily raked transoms; flush decks with paint or moldings to hide the resultant freeboard; and a curious sheerline which, though springy, descended continuously from bow to stern. Falmouth Quay punts were examples of small fishing boats of this type. Of the larger sizes, most respected were the Bristol Channel pilot cutters. All had gaff rig, deep and narrow bodies with much deadrise, and inside ballast.

When steam replaced sail, many of these sloops and cutters were converted to yachts, and proved themselves on long passages. The Pyes sailed a Falmouth punt trans-Atlantic, and wrote about it in *Red Mainsail. Jolie Brise*, a famous racer of the '20s, was a retired pilot cutter. In *Mischief* and her successors, all pilot cutters, Major H. W. Tilman made a series of voyages to extreme northern and southern latitudes, circumnavigating Spitzbergen, making first ascents of slippery mountains, and losing boats in the ice until he himself was lost on a voyage to Antarctica at age 80. Stone deaf, violently opinionated, addicted to scalding curries, he made do with pick-up crews. He wrote eight books about his voyages, and they are among the best and funniest cruising yarns of our times.

Over years, these sailing workboats inspired a number of yacht designs, most prolifically from Lyle Hess. He designed the 23-foot *Seraffyn* and her successor, which the Pardeys have made familiar to all yachtsmen. It should be emphasized that these boats are small in overall length only, and that another *Seraffyn* would cost at least as much to build and own as a modern 30-footer, except when paying dockage. The *Cornish Crabber* is yet another variation on this theme, though more in looks than in form.

I tried to persuade Bobby to make his new boat round bottomed, but he was still feeling loyal to the *June Bug*, and was still convinced by the rhetoric he'd invented to defend her. He wanted a flat-bottomed Bristol Channel pilot cutter. When I told him it wasn't possible—flat-bottomed sailboats must be long and light—he agreed to a V-bottom, but insisted on chines.

Bilge Keels

More reasonably, Bobby wanted shallow draft. On a blow-out tide, his anchorage has as little as a foot of water. We could have used a centerboard as in the Crabber, but this inevitably makes a heavy boat; because the ballast is higher up there must be more of it. Negotiations were going so well that I dared suggest bilge keels, and was delighted with how quickly he took to them.

Because tides of 30 feet and more leave most moorings dry for part of every day, bilge keels or twin keels have been popular in England for many years. Single-keel boats must be propped up with legs, but bilge keelers just sit on their keels until the water returns. Douglas Philips-Birt, in his book *Sailing Yacht Design*, devotes many pages to bilge keels, and feels that once they are perfected, they may prove faster than single keels. Forty years after he wrote, this has not proven true. It appears that two keels, even if smaller than a single keel, must have more drag and must be marginally slower. But their advantages may offset this liability, especially in shallow water.

No one system of designing bilge keels has ever been settled on. Robert Tucker and Maurice Griffiths favor a large center keel which stiffens the structure, gives directional stability, protects the rudder, and houses the ballast. Their bilge keels are small fins that only serve to provide lateral resistance and stability in taking the ground. With the boat upright, all three keels are the same depth. Other designers, such as Robert Clark and Arthur Robb, use no center keel, but use larger and longer bilge keels, making them do all the work. There probably isn't much to choose between these two theories, but I felt that two keels were enough to build and drag through the water, and Bobby agreed.

One great advantage of bilge keels is that they draw more water when the boat is heeled than when it's upright. *Dr. Buzzhead* draws 24 inches upright, but 28 inches when heeled to 15 degrees, which should be her best sailing angle to windward and is the angle of the deadrise of her bottom. By contrast, a single keel draws less when heeled. It seems clear then that bilge keels should angle outward to be vertical when the boat is heeled to her best sailing lines, but this is not always done. The windward keel is doing little for leeway resistance at that point, though it may help dampen roll.

The time when a sailboat is most likely to run aground is when beating to windward, using all of the channel and then some. *Dr. Buzzhead* really comes into her own then. At the first touch of bottom, you put the helm down, the boat heads up and the sails luff. The hull springs upright and draws less. No longer touching bottom, it continues to turn, and eventually flies away on the new tack. This works very well over a sandy or muddy bottom, although I wouldn't try it over rock.

Some have thought that as the leeward keel is providing most of the lateral resistance, toeing in a few degrees would give desirable lift to windward. Downward pressure on the toed-in windward keel would help counteract heeling. Maybe so, but tank tests have shown that the drag of toed-in keels hurts performance on all points of sail more than lift to windward helps. The result is a slower boat. Bilge boards (twin centerboards) are usually toed in a bit.

I judge that *Dr. Buzzhead*'s keels are as effective in preventing leeway as a single keel of similar length (about one-third of waterline) and a foot more draft. Bilge keels are not nearly as effective as a daggerboard or a deep fin keel, but they are better for Bobby's locale and they are well matched to the rig.

Each keel of *Dr. Buzzhead* has a 265-pound lead shoe. Total ballast is therefore only 21 percent of boat weight, and crew weight does as much as ballast to stabilize the boat under sail. She's proportionately beamier than *June Bug* (V-bottoms can have a little more beam than flat ones) and as the cockpit still isn't self-bailing, crew weight is low and well outboard. With a self-bailing cockpit, it's surprising how quickly heeling puts crew weight on centerline, and useless as ballast. Self-bailing certainly has advantages offshore, but the chief advantage inshore is that it allows the boat to be neglected.

Scaling Down

When scaling down, it is usually best to simplify as well. When Bobby's Crabber shrank to 20 feet 7½ inches by 7 feet 4 inches and 2400 pounds boat weight, she lost her topsail and staysail. The Crabber itself is really too small to profit from these sails. Staysails have a place on big boats, but even the behemoths set a single headsail for racing, because one big sail generates more lift than two sails half its size. From what I've seen, double headsails don't draw properly unless the luffs are at least 3½ feet apart from tack to peak, and this observation is based on wind pressure and doesn't scale down for boat size. I doubt that topsails ever earn their keep, unless the topmast can be struck along with the sail.

Dr. Buzzhead's mainsail follows John Leather's recommendation in his book *Gaff Rig* that foot and luff be about equal, and the head about 15 percent shorter. The head is peaked up to about 30 degrees from the vertical, which is a good compromise between leading edge and total area. When you are doodling gaff sails, the total area is often very nearly equal to luff times foot. To get a proper pull with the peak halyard (and this gaff sail is big enough to need two halyards), the mast should be at least a foot higher than the middle of the gaff. In a sail this size, both halyards can be sweated satisfactorily without multipart purchases.

The jib is roller furling, and not as close-winded as a hanked jib. Though anachronistic, a roller-furling jib works better with a gaff main than with a closer-winded Bermudan. The sail is 70 square feet, which is near maximum for a headsail without sheet winches or purchases. I don't like two-part headsail sheets, which usually have blocks that klunk around when you tack, and always result in a big pile of line in the cockpit. Better to buy winches, or else be prepared to take a luff when sheeting in a fresh breeze. I chose a furling gear that would handle a sail almost twice this size according to the manufacturer, and it always worked fault-lessly. For Bobby, who sailed often but only for a few hours, speed of getting under way was especially important, and he liked the roller furling. He kept the mainsail on the spars, but covered against sunlight.

Some sailors may wonder why, in a boat that is usually daysailed, *Dr. Buzzhead* has such a large cabin and such a small cockpit. Many people want boats for their dream use, not for their real use, and Bobby was one of them. He and Gail did spend a night aboard once or twice a season, but usually within a mile of their mooring. Gail is a good-natured woman, and even after marriage, she has remained an interested sailor. But her in-

Figure 5–5. Gail shows off companionway doors and inboard engine.

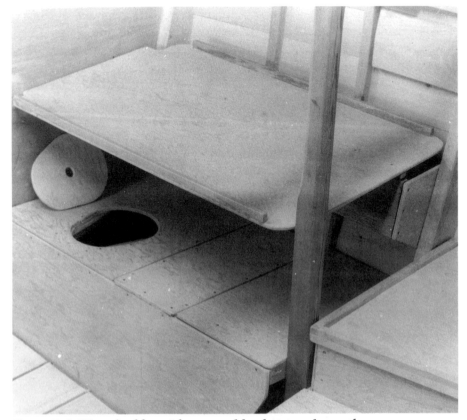

Figure 5–6. Hinged table, with seat and backrest underneath.

terest ends at the inlet, and perhaps, for all his bluster, Bobby's real interest ends there, too.

A double bunk was the first accommodation requirement for the new-
lyweds. The mast was in a tabernacle, and I explained to Bobby that it
needed support, either with a post to the keelson that bisected the bunk
or else with a deep deck beam on which he could easily crack his head.
"My head's pretty hard," he said. "I don't mind cracking it once or twice.
But that post in the middle of the bunk, that ain't so good."

Aft of the bunk, there is standing headroom in a hatchway 3 feet by 2½
feet. A stove and sink are to starboard and, to port, is a seat with bucket
head under it and pivoting table above. This sounds tricky, but it works
very well, and all social and bodily functions are easily accomplished in
Dr. Buzzhead's cabin.

Bobby wondered if this bigger boat (on her marks, she displaces almost
twice as much as *June Bug*) didn't need a bigger engine. Again, the deciding factors were the centrifugal clutch and reduction gear, only available

on the 5-horse Honda. It gives 3.4 horsepower per ton, right in the middle of the "normal" range, according to Bill Durham. The *June Bug* had a 3-bladed prop, 8 inches square, found for a bargain price. Expecting better sailing performance from *Dr. Buzzhead*, I sprung for a 2-bladed sailboat prop, 10 inches by 8 inches. She always powered fine.

Bobby wanted no plywood in the hull of his boat, and the rounded topsides couldn't have been ply in any case. We settled on carvel topsides and cross-strip bottom. This is exactly the reverse of what it should have been. *Puxe* may need a cross-strip bottom to take the pounding of her speed, and because she doesn't heel and immerse a seam, she gets away with carvel cedar topsides. In *Dr. Buzzhead*, a carvel bottom, whether cross or fore-and-aft planked, would have been strong enough and would have swelled and stayed closed. Strip-planked topsides wouldn't have dried out and leaked when the boat heeled.

Still, this gave me some carvel planking to do, and that's the most fun of any boatbuilding work. No one will pay you to smoke cigars or drink whisky, so it's astonishing that anyone will pay you to hang carvel planking. As we were going to varnish the top two strakes and paint below them, this planking had to be spiled, and that added to the fun. Carvel planking goes fast and makes very little sawdust (until the fairing and finishing, but that's another job for another day). Everything about the work is delightful except the end product, which is not strong for its weight, and is prone to work and shrink and leak.

What with Bobby's traditionalist banter and the reasonable prices at local sawmills, some meaty materials went into building *Dr. Buzzhead*. Bow post and stern knee were 4-inch by 6-inch oak. Frames were 1¼-inch oak on about 3-foot centers, with cleats between them so big that, even steamed, they would hardly take the curve. Longitudinals were also 1¼-inch oak. Bottom was cedar, finishing about 1⅛ inches, topsides were ⅞-inch, and decks ¾-inch strip plank. I kidded myself by saying that this wasn't a MORC racer, and that the lightest materials and scantlings wouldn't save more than 300 pounds in a boat of 3400 pounds displacement. That's all true, but 9 percent is a very worthwhile weight saving, and makes a performance difference in any boat. I also wasn't allowing enough displacement for the geegaws that Bobby would add, and that most traditional boatowners like to add. The plans I've drawn for this book show lighter but sturdier scantlings and more glue, and the result has to be a better boat. Some of the weight saved higher up is added to the keel shoes, which are each 100 pounds heavier than originally.

The first geegaw that Bobby bought for the boat was an Edson rack-and-pinion steerer. This heavy unit, suitable for a minesweeper, was all outboard of the waterline stern, and drew the crew weight several feet further

Figure 5–7. Mast-supporting bulkhead has hand-hold (right).

aft than a tiller would have. It was the main reason we eventually had to put 200 pounds of ballast under the foot of the bunk. For the cost of it and its installation (I refused to do it), the boat could have been a foot longer. It worked like butter, but transmitted no feel of the water. Bobby was nuts about it.

Dr. Buzzhead sails well. Her performance numbers are about average for a sailing auxiliary and her rig is somewhat less efficient than a modern one, so she gradually loses ground to a modern boat such as an O'Day 23. But very gradually. Compared to other traditional boats, her performance is startlingly good. In part, this is because most traditional boats are now 50 or more years old, broken-framed, paint-sick, and baggy-sailed. The boat was also a better design than many traditional ones — lighter, simpler, better balanced, and with less wetted surface.

She is not a descendant of the working sailboats of Southwest England in much besides looks. The chine gives her harder bilges, and she has less deadrise, a beamier hull, and outside ballast. Her motion and handling can have little in common with Falmouth Quay punts. In a hard chance, she is probably less seaworthy than they were; but, inshore, she's every way a better boat.

In the two or three years Bobby owned her, *Dr. Buzzhead* lost some of her performance. Eventually, he confessed to me that he was reefing earlier than he once had, and asked me why. I could only point to the way he had festooned the boat. From the start, he insisted on two shrouds each side, but the ratlines didn't go on until the second season. Then there were the giant running-light boards (solid mahogany) that some handy relative made for him. There was a string of flags of all the countries he'd be sailing to someday. Flags have a low lift-to-drag ratio, and he seldom set sail without them. In addition to all this windage, there were the solid brass bell and tilt meter on the aft bulkhead, the teak racks in the cabin for storage of tableware and utensils. All of this weighed something, and none of it was low. Finally Bobby evolved the theory that the heavier the boat was, the better it sailed, and all it needed was more inside ballast. He bought pigs, but before he installed them he traded *Dr. Buzzhead* in on a broken-framed old Alden.

I was sad. A craftsman is paid money for his work, but almost always it is less than he could earn elsewhere. Another part of his reward is a residual interest in his product, and that is why owners of paintings will always lend them back to the artist for a show. Ten years ago, a promoter persuaded dozens of Downeast craftsmen to rally 'round and build him a new coasting schooner, the *John F. Leavitt*. When he sank her on her maiden voyage (he thought he had to conform to the schedule of a filming crew), one craftsman wrote a letter to *National Fisherman*, saying that the wages the promoter had paid them all obliged him to take better care of the boat.

Tuckahoe Catboats

It's amazing that catboats sail at all. The proportions are all wrong for a low-powered boat, but the large, efficient rigs get around that. With all the ballast inside, they shouldn't stand up to those rigs, but great beam (i.e., bad proportions) overcomes that. Catboats with barn-door rudders are arm-numbing to steer, and often even tiny ones have wheel steering with substantial gear reduction. One and all, they head up when enough of the rudder is out of the water. Catboats are often fast for their length, but only the best of them are fast for their displacement.

June Bug, the first boat that I built on speculation, was sold before she was finished, and from the response the advertisements brought, more than one could have been sold. I figured that if wooden sailboats sold that easily, fiberglass ones should sell easier yet, and in the fall I laid up two hulls for the *Tuckahoe Catboat*. This was a mistake. New or nearly new wooden boats are in small demand, but they are also in small supply. If the design is reasonably sound, the advertisement for such a boat always

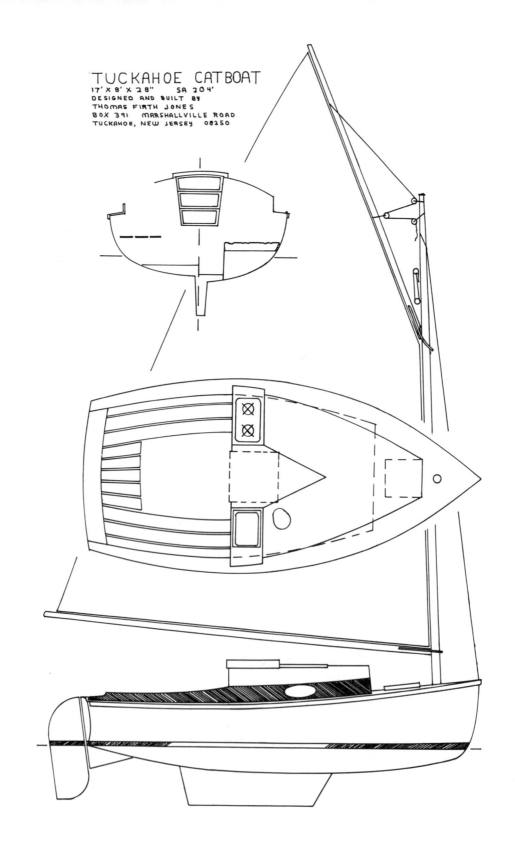

TUCKAHOE CATBOAT
17' X 8' X 28"   SA 204'
DESIGNED AND BUILT BY
THOMAS FIRTH JONES
BOX 391   MARSHALLVILLE ROAD
TUCKAHOE, NEW JERSEY 08250

Figure 5–8. Catboat well heeled, but with plenty of rudder in the water.

brings a flood of inquirers. These people are usually older, and are accustomed to buying craftsman-built products and are not afraid to judge them or hire a surveyor to judge for them. Respondents to ads for fiberglass boats are fewer, and usually flakier. A person who wants such a boat customarily goes down to the fiberglass boat store and picks one he can afford the payments on, or the one that looks good on him. This was true when I first learned it, seven years ago, and it is more true today. There are buyers who appreciate the work of fiberglass craftsmen, but not many.

Fiberglass work is alienating, because no matter what suits are worn and fans are run, it's smelly and filthy and unhealthy. It's possible to lay up a hull or two nonchalantly, but if the work is done steadily, alienation sets in. I've never known a conscientious fiberglass worker, and when op-

Figure 5–9. Catboat moving deceptively well in light air. Circle above water-line is sink drain.

portunity arises, many will go beyond carelessness to sabotage. Urinating in the mold is standard. Beer bottles and pails of uncatalyzed resin are routinely chucked into keels. Whenever possible, the work is done drunk.

Not too long ago, a 50-foot sportfisherman, the product of a prestigious local manufacturer, hit a jetty off the Jersey Coast and sheered off her keel. Analyzed by the underwriters, it was supposed to have a 35-percent glass-to-resin ratio, but in fact had 5 percent. Stories like this are common jokes around boatshops; everybody enjoys them and understands them. A good surveyor who can find any defect in a wood or metal hull is hard put to find the ones in fiberglass hulls that result from worker alienation. I always advise people who want a fiberglass boat to buy one from a small builder who doesn't do fiberglass work all the time. Better yet, buy one from an amateur who built the boat for himself.

Like the wheelbarrow garvey, the catboat hulls were foam sandwich, laid up over a mold of frames and stringers. Especially in a round-bilge boat like this where curves stiffen the glass, there is a point when it's better to make a full mold and save the cost of the Airex. I think this point is four or five hulls, not two. The keels, which were solid glass and were later filled with reinforced concrete, should have been laid up separately on a bench, and glassed to the hulls later. I laid them up on the hulls, however, where they were hard to reach. The bulkheads, also foam sandwich, were laid up on a table and taped to the hulls with V-shaped foam fillets to avoid hard spots.

I used no plywood anywhere in the catboats. I thought this touch of purity would appeal to buyers who had already compromised themselves with fiberglass hulls. It may have, but it added 100 pounds to each boat. Much of the wood was varnished (cabin sides, cockpit seats, the whole interior), and however nice varnished ply may look, varnished lumber looks nicer. Decks were strip-planked cedar, and cabintops were cedar

Figure 5–10. Catboat mold.

boards, polyurethane caulked, fanning out from forward aft over laminated beams. They looked pretty, but they leaked later, just like carvel topsides (I hope the owners have now canvassed them). Boards come and go too much to be glass covered, unless the glass is very thick.

A strong deck beam was at the forward end of the cabin. Between it and the transom, one long mahogany board formed the cockpit sides and coaming, and was the whole structure. Decks were hung on a ledger that was glued to it, and cabintop beams were notched into it. It was stiffened in the middle by the main bulkhead. It was very strong, and has given no trouble.

The concrete keels did not put the ballast as low as lead would have, but they were one-third of boat weight, yielding a center of gravity a lot lower than would have been possible with inside ballast. They also allowed a rudder deep enough to steer the boat, which could then be a decent aspect ratio, and not too hard to handle with a tiller. The lower center of gravity allowed the boats to be narrower on the waterline than most catboats, so better proportioned. One might argue that they aren't really catboats, but they did look like catboats, and had the usual catboat rig.

Performance

These rigs are powerful, as only single-sail rigs can be. The 1.52 aspect ratio doesn't look high, but it must be compared to an entire sloop rig, not to the individual sails, because in sloop rigs the work of the two sails is interconnected. Twist is not so severe in a catboat sail as in a lower-peaked gaff. The gaff jaws have no shrouds to chafe on, as they do on stayed gaff masts.

When we launched the Tuckahoe Catboats—Bobby Graham borrowed a crane, drove it up, and handled it as carefully as a motorboat—we thought they were pretty slow. They made no noise and not much wake, and it wasn't until we took one down river and tried her against other boats her size that we realized she wasn't slow at all. The catboats were sneaky fast, moving through the water faster than they seemed to be going and faster than they looked like they would. You can have some fun with a boat like that.

Coming home from a boat show in Philadelphia, down Delaware Bay, I passed the mouth of the Cohansey River and saw an average-looking 25-footer set out after me. I was beating, and thought I still had the tide. When it became clear that I didn't, I turned back for the Cohansey, and tied up for the night in a Greenwich marina. The skipper of the 25-footer came over after a while. "Upwind or down, I couldn't get anywhere near you," he said. "What's the explanation?"

Figure 5–11. Fin-keel catboat in slings.

I had a weary time selling the catboats, not having anticipated the idle shopping habits of most buyers of fiberglass boats, or the number that would expect to trailer a cruising boat with an unstayed rig. Customer after customer said they liked everything about the boats except the keels. I pointed out that the keel was the whole reason the boat was so fast and had such nice accommodation. Never mind. The customers had a mania for the mechanics of trailering. It interested them far more than the mechanics of sailing. One bozo spent a whole afternoon with me, trying to figure out how the deck beams and cabintop could be reinforced if he cut them so that the mast could be raised and lowered. I was relieved when he departed.

The public interest in trailerable sailboats has now declined (along with most other public interest in sailing). The Groots (see *Scampi*, Chapter 6), who came to me wanting a trailerable cruiser and not sure whether it should be sail or power, decided on power, and they were right. A wide variety of boats can be trailered home once a year, but if getting under way can't be quick, people just won't use a boat after the first season or two.

Selling the catboats, I sailed nearly a thousand miles in one or the other of them. Many of those miles were singlehanded, and I learned that catboats are as easy to sail as the advertisements claim, provided the wind is steady. In squalls, they're not so handy. In shallow water, they can be anchored for reefing, but in deep water, there is nothing to hold the boat up into the wind while the single sail is reefed. Neither boat was rigged with all the strings leading to the cockpit. This would have required a number of fairleads, but might have been worth it. To get the halyards as taut as can be done by sweating, halyard winches would have been needed.

Beam, rather than length, gives the feeling that a boat is seaworthy and will stand up to strong winds. But length, rather than beam, is what actually allows a boat to keep going in reefing weather—especially if the wind is forward of the beam. The catboats balanced well, whether single or double reefed, as well as under plain sail, but there was a point when they stopped making progress. Once, motoring out of the C & D Canal into Delaware Bay, we faced perhaps 30 knots of wind slightly forward of the beam. The sail, with a double reef tied in, was immediately raised, but the boat just wouldn't go and we were glad to retire behind Reedy Island to wait it out. I doubt that other boats 16 feet on the waterline would have done better.

The two people who eventually bought the Tuckahoe Catboats have treated them decently. Ron Kushner loved the varnish, and looked forward to getting at it each spring. Although it isn't my idea of fun, there's no reason why a pleasure-boat owner shouldn't find as much pleasure in varnishing as in sailing. After a few seasons, Ron decided he wanted not just a good varnish job, but a perfect one. Instead of three coats, he put on seven. He and his girlfriend were at it until August, and he told me the finish was so deep and lustrous that it had been worth every hour he'd spent. But the next spring, when he saw that it needed doing again, he sold the boat. I told the new owner that he could paint any part of the boat he wanted, and it would still be a nice boat. No, he said, he liked the varnish. He was thinking about seven coats for next year.

Mock Turtle

The first boat I built in my new shop in 1986 was the first job with a tight enough schedule to require hiring someone to work with me. Hard to say what pleased me more: the shop, the commission, or the company of Frank Blizzard. Working alone, you learn what you can from reading and from the materials, but Frank had plenty to teach me. We had been carpentry foremen together at Yank Boats, before the bankruptcy, and now Frank quit a house-carpenter's job paying several more dollars an hour, because he wanted to build another boat.

A bear of a man, with a thick black beard and thinning hair, neither of which he ever cuts, Frank is a wonderful problem solver, curious and interested in every phase of the work. His giant hands can disassemble the tiny parts of the sealed switch in an electric drill and make it work again. In addition to carpentry, he did all the mechanical and electrical work on *Mock Turtle*, and I never had to check on him.

The new shop is more or less the same 600 square feet as the old, but it is high-ceilinged and well lit, both naturally and artificially. Under the house, it needs little heat, and is cool in summer. The doors, 9 feet high and 10 feet wide, can handle the biggest job I ever want to tackle. In a shop so small, where boats of various sizes are built in various places, the most important feature of stationary tools is that they be portable. I have very few—table saw, band saw, jointer, drill press, grinder. They have given good service, some for 30 years, despite being the cheapest available. My power hand tools are the best that money can buy, on the other hand, and many of them only last two or three years. In unpowered hand tools, I enjoy using the best, but am not sure they do better work.

Mock Turtle was designed by her owner, K.V. Weisbrod. Nearly 70 years old but vigorous and a life-long yachtsman, he had owned a Robert Tucker plywood 35-footer, and more recently a Primrose-designed 30-foot strip-planked sloop. Now he wanted a smaller boat (the idea that you should have a bigger boat as you get older is only sensible if you have someone to do the work for you), and he remembered the Tucker with most affection. He drew a single-chine, 24-foot plywood yawl, with flush deck and bewilderingly complex layout. I told him I could fair up the hull lines, develop the sections, and figure the weights and centers, but he went to Duncan MacLane, best known for his class C catamarans. Duncan drew the lines with the lengths to one scale, and the breadths and heights to another. This is often done with multihulls because the curves are so gradual that you may otherwise miss an unfairness. With *Mock Turtle*, it produced a lines drawing almost as high and wide as it was long, and very curious to see.

Duncan's work was careful, and the planking went around the developed sections exactly as predicted. Often, plywood boats this size have stringers over regularly spaced bulkheads, but in *Mock Turtle* (as in the plywood Chris Crafts of 30 years ago) there are no stringers, and the skin is most of the structure. We built the hull over molds, and spiled in most of the bulkheads later. In planking a carvel or clinker boat, scribing is often quicker and easier than spiling; but in fitting bulkheads, spiling is wonderfully easy and precise.

We built with okume plywood, ³/₈-inch 7-ply for the topsides and ¹/₂-inch 9-ply for the bottom. A more rot-resistant species such as khaya or

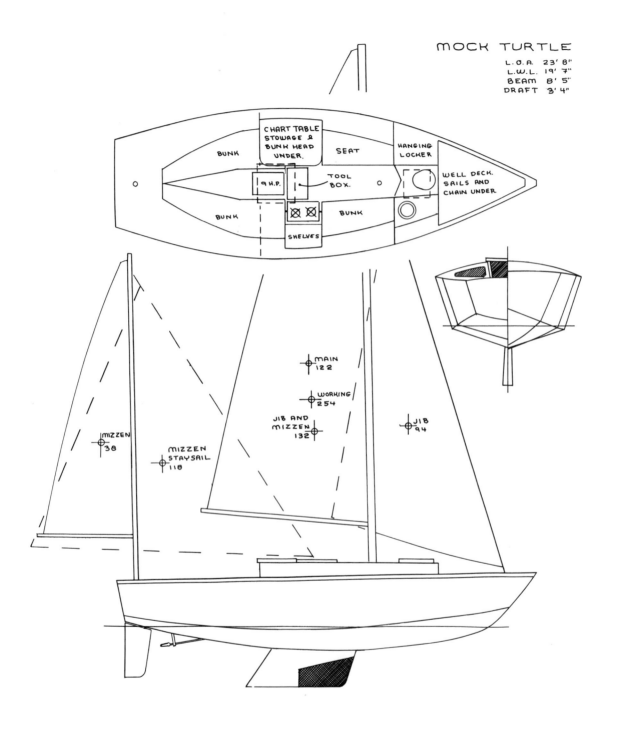

MOCK TURTLE

L.O.A. 23' 8"
L.W.L. 19' 7"
BEAM 8' 5"
DRAFT 3' 4"

BUNK

CHART TABLE
STOWAGE &
BUNK HEAD
UNDER.

SEAT

HANGING
LOCKER

9 H.P.

TOOL
BOX.

WELL DECK.
SAILS AND
CHAIN UNDER

BUNK

BUNK

SHELVES

MAIN
122

WORKING
254

MIZZEN
38

MIZZEN
STAYSAIL
118

JIB AND
MIZZEN
132

JIB
94

Figure 5–12. Mock Turtle *on the Tuckahoe. Blob amidships is tricky chain-plate.*

sapele might not have taken the twist of the bottom. To protect the okume from rot, we sealed all end grain carefully with epoxy, and painted the inside with latex paint.

Bill Weisbrod came down to help us sheathe the hull with Cascover. He was a little startled by Frank's beard and the aggressive logos on his tee-shirts, but was soon reassured by his light tenor voice and his unfailing courtesy and competence. Cascover had been on Bill's last two boats, and by God it was going to be on this one. He said he's seen holed boats sail back to port with nothing to keep the water out but Cascover sheathing, and it sure is stronger and more flexible than fiberglass. He imported it from England where he had spent much of his sailing life and where it is often used.

Cascover is a very heavy nylon cloth of superb quality, stuck down with resorcinol glue. The glue sticks wonderfully to the wood, but less well to the nylon, and this is especially a problem at the laps. The nylon can't be sanded — it comes up in an impossible fuzz — so tailoring must be careful, but the cloth drapes better than glass cloth. The weave is supposedly filled with layers of vinyl paint, but we never did fill it before the paint ran out. I think that Cascover fabric in epoxy resin would make a plywood boat nearly bulletproof. Unfortunately, the fabric and glue and paint are sold only as a complete package by the company that makes the glue. On *Mock Turtle*, the dried Cascover showed blisters here and there, and Ruth Wharram tells me they've had the same trouble with it. We slit the blisters, injected them with epoxy, and pressed them down until the glue went off.

Bill's layout packed at least as many features below as he'd had on his 30-footer: diesel engine, three bunks and a settee, stove, large chart table with drawer under, enclosed head, hanging locker, chain and sail lockers. Many of these features overlapped in tricky ways. In one bunk, the sleeper's feet were supposed to go under the sink in the head. Bill couldn't tell us how to make this work so we left it out, but Bill says he still plans to figure it out and make it work. Big Frank experimented, and the only way he could sit on the head with the door shut was to close the door from the cabin, go on deck, and drop down onto the head through the forward hatch. Bill is smaller than Frank.

The yawl rig is also not the simplest to handle from a platform 24 feet long. But Bill honestly admits that he likes puzzles, and before he settled for a hasp and padlock on the companionway, he was many weeks thinking about a secret place where a string could be pulled that would actuate a lever that would. . . . I forget how it all went. In a pleasure boat, the critic should think twice before criticizing how the owner gets his pleasure. Complexity, however, is more expensive than simplicity.

For Bill's other interest, I have much more sympathy: He cares more how things *work* than how they *look*. The preoccupation with looks in functional objects (such as human bodies) is the greatest modern absurdity, and probably comes from television advertising with its incessant appeals to our vanity. It is the modern absorption with finish instead of structure. Bill Weisbrod didn't necessarily want an ugly boat, and he didn't get one. But he had so little patience with aesthetics that I was some time persuading him he must have a curved bowpost instead of the straight one he first drew, because a straight one couldn't be planked in plywood. He did finally agree to this, and he at once recognized that a tapered washboard would come out of the companionway easier than the rectangular one he'd drawn. Aside from that, he was not receptive to suggestions for changes, especially if he thought they were for looks. Bravo, Bill!

The single-cylinder diesel did fit under the cockpit sole, though without much room to work on it. Being Japanese, it hasn't needed much work. It turned a 2-bladed folding propeller. I cut a mark on the coupling to show when the prop would be most likely to fold (i.e., when the hinge pins were vertical), but I doubt that Bill crawls in there to look at it very often. Under sail, he must usually go along with one blade folded. I don't think much of folding props, which often foul and then won't open or won't shut, but it was a gimmick and Bill liked it. In forward or reverse, it pushed the boat okay when everything was new.

Mock Turtle also had an Edson steerer, this one operating chain over a sprocket. On this little boat, there wasn't enough vertical clearance under the steerer for the link between chain and cable. With the rudder fully turned, the link jammed either the sprocket or a cable sheave. A drum had to be machined, and the cable wrapped around it, and the stainless steel chain and sprocket set aside. Buying these contraptions and getting them to work on small boats costs the world—often more than the boat itself.

The high aspect ratio rudder was very powerful. One might think that transom-hung rudders, not being end-stopped, would work no better than leeboards. But the water passing over a transom-hung rudder is already much disturbed by the hull, and the rudders work acceptably well. I prefer them for their ease of installation and maintenance. But for pure steering power, inboard rudders are best.

The pattern for the ballast of the fin keel (1100 pounds of lead) was made from Duncan's elaborate offsets. The fin itself was a standard foil section, which meant that it tapered, and was thickest at the bottom and thinnest at the top. My foundryman said it was very good to have a ballast casting come within 5 percent of its designed weight, but this one came back exact to the pound—a tribute to Duncan's math. *Mock Turtle* was an odd mix of simple ideas (like the single chine) that Bill Weisbrod had found to work, and Duncan MacLane's ultrasophisticated specifications. As I was building the boat for a contracted price, I soon learned that when something was left out or not clear in the drawings, it was cheaper to talk to Bill than to Duncan.

Bill sailed *Mock Turtle* out of here. For crew, he had a college classmate who was a foot taller than he and a hundred pounds heavier. Before leaving, they spent a couple of nights aboard the boat at dockside, and Will's expression became grimmer each time we saw him. We learned he was having trouble sleeping in the 22-inch bunk he'd been assigned, under the cockpit. We didn't dare ask if he was managing the head any better than Frank had. However, he survived the trip to Long Island, and he and Bill are still friends.

Experiments

Bill Weisbrod regards a boat as a platform for experiments, and he lacks neither ideas nor energy. Periodically, I would hear from him that he had replaced the companionway ladder with a tool box, installed a larger mizzen, moved the keel forward 13 inches, etc. Last I heard he was talking about a leeboard to balance the mizzen staysail. I didn't encourage him.

After the America's Cup was won by a catamaran, Bill called up and said, "I think I've got the wrong boat!" He began drawing a catamaran, and we exchanged a good many letters over it. I was fascinated to see the same mind working on a different problem. He immediately grasped the necessity of light weight, and kept the deck flush and the accommodation spare. But *Yum Yum* had a nightmare steering system, with small quadrants attached to the transom-hung rudders below deck level. From the quadrants, cables fed through water traps in the transoms, ran along the overhead of the aft bunks, up through the aft crossbeam, and to a wheel steerer amidships. In the end, money kept Bill from going ahead with her. The complexities of the design pushed my bid higher than the original estimate, and I suspect he found that the boat he had, like most sailboats, wasn't worth much on the market. Though I would have enjoyed building him another boat, he made the right decision. For a man in his 70s, *Mock Turtle* still offers many opportunities for experiment. And for a small cruising boat, she sails very well.

Carol and I visited the Weisbrods during his boat's fourth season in the water. We daysailed her, and it was a pleasure. Aboard, we could see that Bill was a little behind on cosmetic maintenance, but only a little. He hadn't loaded her down with junk. Except for 50 pounds of tools, I doubt that *Mock Turtle* weighed any more than when she left the shop. He showed us the improvements. She sure had plenty of running rigging, especially for the mizzen, which he is determined to sheet without a boomkin.

The diesel wasn't working that day—even the Japanese take a day off once in a while—and the harbor was fearfully crowded. Bill fussed around like some of our *El Toro* competitors, but before too long we were under way. Slipping between the moorings in very light air, Bill read wind and current astutely, and his moves were deliberate and relaxed.

In Long Island Sound, with somewhat more wind, we beat against it, and I had to admit that *Mock Turtle* sailed more crisply than the one other time I'd tried her in the Tuckahoe with the old mizzen and old keel position. "Not bad for a chined boat, eh?" Bill said. In truth, she sails well for a 24-footer of any hull shape. I doubt she'd stay with a J-24, but she has the same low slug of lead, and the same high narrow sails, and the result is impressive.

Off the wind, Bill wanted to show us all the sails, one after the other, but we settled for a big light genoa and the mizzen staysail, because Carol had never sailed with one before. The boat did certainly gain a knot when we set it. Like topsails, I doubt that mizzens earn their keep, upwind and down; but they are fun to use, and sailing is supposed to be fun. That day we had a lot of fun with Bill on *Mock Turtle*.

Sailing Pocket Cruisers

6. Power Pocket Cruisers

Johan and Meit Groot are semi-retired geologists, born and educated in Holland, who became American citizens early in their working lives. Off and on they have taught at the University of Delaware, but have also prospected for water for the U.N. in Bolivia and Chad, and spent a good many years in England. There they became friends with Blondie Hasler, and bought two of his junk-rigged sloops, the first one a delight and the second a turkey, according to Johan. In a more conventional 25-foot sloop, they made a tradewind crossing from England to Barbados. Their last boat, sold a year or two before they came to see me, was a steel motor cruiser, reliable but wet.

The Groots knew their requirements: a cruising boat that could be easily trailered, and would get them away from the heat and familiarity of the Chesapeake in summer, perhaps to Maine. They didn't know what type of boat they would enjoy most, so they brought along two sets of pictures. One was a flat-bottom plywood cat yawl with relatively modest accommodations, and the other was Sam Devlin's *Surf Scoter*, a 22-foot semi-planing outboard cruiser. They were worried that the yawl might be too cramped and too slow, but were sure the Devlin would be too big and heavy for them to trailer.

Without hesitation, I recommended the powerboat. The yawl was lug-rigged, and I doubted it would sail well. The Groots may have put up with a junk rig 20 years before, but most of us get less patient as we grow older, and John made little secret that this was happening to him. The cramped cabin, perhaps tolerable in the Chesapeake, would have been a weary place on the cold, foggy days in Maine. The sailing rig would make hauling and launching the yawl at least as much of a project as the heavier Devlin boat. Besides, what about scaling down the *Surf Scoter*?

Scampi

The Groots, too, had thought about scaling down, but weren't sure how to do it. They didn't need 22 feet of length or 6 feet 4 inches of headroom, and they didn't want 2200 pounds of weight. How to begin?

Sam Devlin's boats are hard to pin down, because they are made of taped seam plywood, so he's not locked into tooling the way fiberglass builders are and he keeps tinkering with the designs. In addition to build-

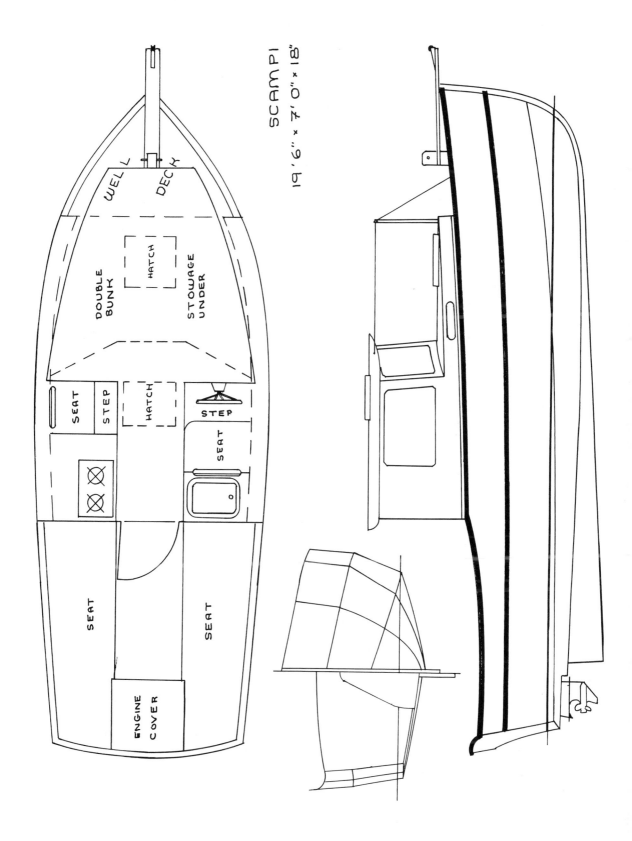

SCAMPI

19'6" x 7'0" x 18"

WELL DECK

DOUBLE BUNK

HATCH

STOWAGE UNDER

SEAT STEP

HATCH

STEP

SEAT

SEAT

SEAT

ENGINE COVER

ing, Devlin sells plans for a range of sail and power boats. I have only seen his work in photographs, but people say his materials and craftsmanship are good.

Devlin has built up a good business because he understands the appearance that many people want in a boat. Though his materials are modern, his aesthetics verge on the quaint. In sailboats, I don't like this. The looks compromise the function, and the force of the wind is so slight (compared to the force of diesel oil or gasoline) that the hardest part of sailboat design is to get performance interesting enough to keep the skipper from giving up and turning on the auxiliary.

Powerboats are a different matter. Even 5 horsepower per ton gives a cruising speed near hull speed, no matter what the hull shape or the windage of squared-off cabins. The hardest part of powerboat design is to make her cute, handsome, or pleasing in appearance—anything to remind the skipper that he's in a boat rather than a car and ought to be enjoying himself. While sailboats are often frustrating, powerboats are often boring. Devlin's quaintness is certainly the best answer to relieving the boredom.

His powerboats have the broken sheer that was used on most cruisers in the 1920s. Besides linking the boats to the past, this lowers apparent freeboard, especially when a second guard is put on about half-way between true sheer and waterline. Transoms are raked and curved. Cabins have slightly raked windshields, overhanging tops (tough to build in fiberglass, but easy in plywood), and big rectangular portlights. Often Devlin puts a round portlight aft of the cabin-side window, but I find this too quaint, like the hexagonal window that you now see on so many houses.

Devlin's powerboats have various layouts—he really does like to tinker here—but all of them center around the steering station. You steer these boats from inside like a workboat, and you take it seriously. You have controls, gauges, switches, and a comfortable place to sit. Someone brings you coffee, does the chartwork, and calculates the E.T.A. You're the captain, even if you don't wear a hat, and you steer the boat.

Below water, Devlin's boats have a long keel, but are otherwise the usual warped-V planing hull with no rocker in the chine aft. This is the same hull we've been seeing since the invention of waterproof plywood, in hundreds of yellowing magazines from hundreds of designers. It isn't a terrible hull. It's less power-efficient than a flat bottom, but quieter and softer riding, and many times more efficient than a deep-V. This hull depends on a lifting surface aft to get it up over its own bow wave, and much of that lifting surface is lost in Devlin's smaller boats by putting the outboard in a well.

Figure 6–1. Scampi *on a trailer.*

Devlin specifies 60 horsepower for *Surf Scoter* which, even with the well cut-out, is enough to drive the boat beyond hull speed. Like a Trojan or an Egg Harbor cabin cruiser, it would trundle along with its bow in the air, but that wouldn't give the desired quaint look and feel, so she has 400 pounds of water in a bow tank. The bow does not rise, and it's too blunt to

Figure 6–2. Scampi *in the Chester River.*

Figure 6–3. Scampi *at dockside. Note Meit's hand on lifeline.*

cut through the bow wave as a semi-displacement hull would do. It ploughs through the bow wave, and moves at more than hull speed. This is not a good way to use power, but it does work.

I suggested to the Groots that we start with a 10-percent scale-down. Theoretically, the weight would be reduced by the cube of 10 percent, or 600 pounds, but this doesn't always work out in practice because it's hard to reduce everything, especially scantlings.

Taped-seam plywood needs to be thick, because it has no reinforcement. I reasoned that if we built over stringers and frames, the way light aircraft used to be built, we could reduce Devlin's bottom from ³/4 inch to ¹/2 inch, and his topsides from ¹/2 inch to ³/8 inch. We'd do without the water ballast forward, and see how she trimmed. We might hope for a 1300-pound boat. Could we possibly put the outboard on the transom, I wondered?

"No!" said John Groot, "We could *not* put the outboard on the transom. It would spoil the whole boat; it would look like an outboard boat." Once again, when *Scampi* was in frame, I asked John if we could put the outboard on the transom, and tried to beguile him with another knot or two of cruising speed, or a smaller engine, or both. He was still adamant. In addition to improving performance, a transom-mounted outboard would have increased cockpit space, and I don't think it would have spoiled the shippy feel of the boat for the crew. Even from outside, there is a very small angle of vision between the point where any boat might be an inboard and the point where you see *Scampi*'s transom cut-out and the charade is over. It can't be more than 40 degrees each side. That 40 degrees mattered very much to John, so *Scampi*'s outboard is in a well.

Designer's Rights

We had a couple of photos of *Surf Scoter*, and a drawing from a design catalog. It didn't seem necessary to buy plans, especially to the Groots, who pride themselves on their Dutch thrift. I don't know Devlin, but this is a hard pill for some designers to swallow, and he may be one of them. Probably the rights of a yacht designer should be as well protected by law an an inventor's or a writer's; but in fact, they aren't. Some designers sell plans with the stipulation that only one boat be built, or that the plans are rented, not sold, to the builder. This is all nonsense, and has no basis in law. When I was drafting plans of the plywood *Tanenui* for James Wharram, I noticed that all his other plans said "copyright" on them, and asked Ruth if you could really copyright yacht plans in England.

"No," she said, "but put it on there anyway. It might stop somebody."

You cannot make and sell a photographic reproduction of a designer's work. But you can make a tracing of it and use it any way you like, because it isn't the same drawing. A designer has only two remedies: He can produce new designs faster than people can copy them, or he can enjoy the mess that most people get into when they try to build one of his boats without plans.

At first, it seems easy enough to borrow from another designer like Devlin and not acknowledge the debt. Sooner or later, this kind of dishonesty will come back and bite you. I was at a boat show in Philadelphia some 15 years ago, and saw a drawing of a heavy, double-ended sailboat. To the entrepreneur in the booth I said, "It's like a Colin Archer, isn't it?"

"No," he said. "I know Colin, and we respect each other's work, but this boat is my design."

If the Groots had wanted a standard *Surf Scoter*, I would not have considered building it without Devlin's plans. Tyros love to save money on hull materials and plans. But even to the most experienced builder, the information in the designer's drawings is worth many times their cost, which is about one percent of the value of the completed boat. Since I was preparing to build a different-sized boat with a different method, however, not much of Devlin's information would have been useful.

From the catalog drawing and from my experience with other plywood hulls, I took a set of offsets at five stations. I scaled them down 10 percent for *Scampi,* and worked them up on the drafting board into a nine-station lines drawing. The only substantial difference between this hull and a 1945 plywood runabout from *Popular Mechanix* was the vertical bowpost, giving a longer waterline and a quainter look. Clearly this would call for a different apex for the plywood development. On the drafting board, I couldn't find it.

Hunting for the apex of a cone is the most aggravating drafting work I know. Draftsmen who do it often may have evolved rules of thumb, but I haven't found many. Chine lines can be moved in plan and profile, and keel line in profile, but there are limits, especially if a certain look and performance are desired. The apex can be anywhere. A very old magazine article sent to me by a friend says that in a planing hull the apex will usually be about 20 percent LOA forward of the chine-stem intersection, one whole beam away from the keel and across it, and 10 to 15 percent below the mean chine height. I couldn't make this work for *Scampi.*

I made a ¼-scale half model of the lower topsides and the bottom from the bow to midships, aft of which the sections wouldn't need developing. The waterline and keel line were drawn on a plywood strongback about 1 foot by 3 feet, and four station molds were attached by their centerlines, as if the half model were on its side. The model was planked with ⅛-inch plywood. This showed exactly what adjustment was needed in the chine line, and the molds were modified accordingly.

Lines were drawn where the bulkheads (which didn't coincide with the stations) would be needed. The model was planked again, and the bulkheads were spiled. From them, it was easy to scale up to the full-size bulkheads, and when it came to planking, *Scampi* gave no trouble. She has conical sections, but where the apex is, I still don't know. Aft of amidships, her bottom panels have a slight arc, 1 inch in 48 inches. It's possible the plywood would rather lie flat, but it's easily tortured this little bit, and it's much stronger convex than flat.

Building the Hull

Scampi went together easily. The keel and stempost were laminated on the lofting from two layers of ⁵⁄₄-inch mahogany. The bulkheads were set up on a strongback, and notched for the keel. The rabbeted pieces between the frames were cut on the band saw and glued and screwed to the keel. In these days of waterproof glues, it's often better to laminate a rabbet than to cut one out; this can save both time and wood. Upper and lower chine logs were also laminated. They are stronger that way, and the bends were easy. Planking was all butt blocked. Some builders like to scarf their plywood planking, and it would have saved a little weight in *Scampi,* say 10 or 15 pounds. It's hard to scarf plywood on the boat without making a lump. If scarfed off the boat, there is often much waste of plywood, and the resulting sheet is hard to handle, especially if working alone. Butt blocks can be made neat, and are certainly as strong as scarfs.

For those who need a really deep-throated clamp for scarfing plywood or any other purpose (they are often handy in clinker work), Jorgensen now sells the hardware for their handscrews in kit form, and you can make the jaws yourself to any length. I made four of them: two 4-footers with 2-foot throats and two 3-footers with 18-inch throats. Douglas fir jaws make lighter clamps than hardwood jaws. Conventional handscrews can be opened or closed by grasping a handle in each hand and flipping the jaws over them, but this is not recommended with the 4-footers. Jaws three or four times the usual length give great mechanical advantage, and by turning the outer screw vigorously, you could undoubtedly break the clamp. But with a little care, these clamps will do many jobs, and I use mine often.

Scampi has no fiberglass sheathing anywhere. For her frequent long-distance trailering, the keel is protected by a 1/8-inch–thick stainless steel shoe. Fiberglass below the waterline may be a useful teredo guard, but this is a trailered boat. The plywood is khaya throughout, which will not check, delaminate, or rot. At first sight it may seem expensive, but the plywood bill was less than 10 percent of the cost of the boat. Once again, the place to save money is not in the structure.

I turned the hull over by myself. Unless the help is as good as Frank Blizzard, I'd rather not have help with a job like this. I usually don't know what's going to happen next, and would rather take my time and not explain. A dependable come-along made from cast metal, not sheet metal, is a very great help. So are exposed ceiling joists in the shop. Holes can be drilled, and all manner of things attached.

The layout of *Scampi* would not suit everyone, but that's one reason for not buying a stock boat. Originally I drew the Groots an asymmetrical

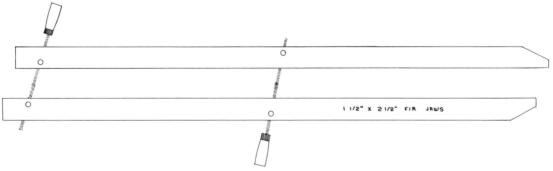

Figure 6–4. Four-foot handscrew with 2-foot throat.

double bunk 4½ feet wide at the head, with a locker on one side, like the bunk in *Dr. Buzzhead*. They insisted on a bunk the width of the hull, 5 feet 8 inches at the head. It does give each of them a backrest for reading before sleep, but 12-volt reading has its limits. Hammocks under the side decks hold clothes.

The pilothouse is 4 feet long, and a comfortable steering station takes up all but 15 inches of one side. I'd have gotten a stove in behind it somehow, and had a forward-facing seat and table on the other side. They put a sink behind the helmsman, with the stove opposite it, and a fold-down table over the stove for chart work. Meit sits forward, facing athwartships. In the bench seat in a sailboat cockpit, you can turn your body 45 degrees, but where Meit sits she must turn her neck to look forward. She is a cheerful and sturdy woman, and hasn't complained. In fact, we think she likes the boat better than John does, because it is so compact and manageable. The bucket head is also in the pilothouse, under the passenger seat. The large storage space under the cockpit sole is one of the advantages of outboard power. You reach into it from the cabin. The Groots keep a screen door in it which can be hung on the same pins as the solid door. They also use this area for stowing dinghy oars, a boat hook, and such.

Their cockpit is pure sailboat, with a bench seat down each side. Much of the space under the seats is taken up with tanks, battery, and the propane bottle. The Groots don't like cockpit awnings. To all Northern Europeans, the sun is a rare treat. Decks beside the house are a minimal 6 inches. Forward of the pilothouse, *Surf Scoter* has only a cambered deck, but we put a low house on the foredeck of the scaled-down boat to get bunk headroom without excessive deck camber. From an earlier boat of theirs, John took the idea of a lifeline from each corner of the pilothouse to a stanchion at the forward end of the foredeck. The hand slides along it without thinking, applying a few pounds of pressure to keep balance until the security of the well deck is reached. Steering the boat, you hardly notice it.

My brother, an architect, designed us a tiny summer house many years ago. To have it look like anything, he said, the details needed to be disproportionately large. It shouldn't look like a little house, but like a wing whacked off a big house. This is at least as true of small boats where charm or character are aimed for. There must be a Samson post, and it can hardly be too large. I'd have liked *Scampi* with a mast and big sidelights, but John wouldn't go for it.

I had hoped the Groots would settle for the 10-horsepower Yamaha or the 15-horse Honda engine. Both are 4-strokes, and marvelously quiet and fuel-efficient. The longer the trips you take, the more it matters. John decided that neither engine would drive *Scampi* fast enough, and bought a 30-horse Yamaha 2-stroke. It's pretty high-tech for an engine that size — three cylinders, each with its own carburetor, and oil injection. It has given no trouble.

The Groots have calculated their speed over a measured course. At 4000 rpm, *Scampi* does 9 knots, or a speed-length ratio of 2.1. I'd guess she's burning about 3 gallons an hour then. They generally cruise at 3500, however, which is closer to 8 knots. The bow is up a few degrees, and it would cost either speed or money to push it down. It would also steepen the bow wave. A quaint boat like this with her bow up doesn't look traditional or charming. It looks like Grandma on her Harley-Davidson, and the only sensible reaction is to get out of the way. So the Groots get most of their compliments about the boat when she is dockside.

They think nothing of trailering *Scampi* to Maine behind their Volvo station wagon. I've ridden with them in the wagon on shorter trips, and it doesn't seem too bad. Driving behind them is another matter. They look like house movers going down the road at 50 mph, clipping off branches and telephone wires. Trailering boats smaller than this makes me nervous, but trailering even bigger boats has become standard practice in America. Horror stories about hitches and tires and wheel bearings are routinely shrugged off. I find the highways scary enough in a car, and don't need to be racing to stay ahead of my own trailer.

We gave *Scampi* her shakedown from a launching ramp at a Cape May marina. Twenty years ago, you'd have seen quite a few clinker runabouts there, and perhaps a faded but elegant displacement power cruiser or two. Now there was nothing but metalflake, over-engined plastic snarlers, and their owners were not the gentle fishermen who favor bridge abutments or tin skiffs, but the hard-nosed, meat-hunting type. Yet several of them came over, and they were genuinely admiring. Seeing *Scampi* made them nostalgic for something, but they couldn't remember what. Maybe it was boats.

In Maine, the Groots say, the wooden-boat folks flock around, and they start so many conversations that it's hard to get away from the dock. People appreciate *Scampi*, whether or not they know anything about boats. She looks proven, trustworthy, and nautical. Despite her high power and boxy shape, she does not look like an automobile. Sam Devlin, whose idea she was, deserves the credit.

Elegant Slider

The name for this design is borrowed from a species of small, green, aquatic turtle with red markings on the sides of the head. These attractive and inoffensive creatures were once widely sold in five-and-dimes, their shells painted with kitschy mottos. The paint kept the turtle's shell from breathing and guaranteed an early death, and after people had known that for some dozens of years, the practice was forbidden.

Though *Scampi* turned out well, I couldn't help thinking how much better she'd have been with a more efficient engine and a lower-resistance hull. Tucked away in a drawer were plans for the Elco 26, bought several years earlier from Mary Farmer, Weston Farmer's widow. Some boat plans are cheap enough to be bought like books, read like books, and stored like books. I've bought quite a few that I never intended to build.

The Elco 26 is a genuine semi-displacement hull. She's about as beamy as such a hull can be—3.3 to one on waterline—and still work. *Scampi* is not much wider, but the turn of the bilge is high in the Elco, and the wave train must make her considerably narrower amidships at cruising speed. The waterline beam at rest gives her the stability to have accommodation in a short length.

The 26 was introduced in 1925 and was successfully marketed for some years, until cheap engines and fuel drove Elco to chined semi-planing hulls in the 1930s. The fire at the Bayonne plant in 1948 may have consumed the company's old drawings—they didn't care much by then, because most of their work was in New London, building submarines. Weston Farmer had drafted for Elco in the '30s and, as he delicately puts it, "managed to secure blueprints" of many of their models.

Like the Concordia yawl and many other designs now thought of as priceless works of art, the Elco 26 was intended to be a standardized, inexpensive boat. Cedar planked, she was fastened with galvanized screws with puttied heads. Running gear and rudder were bronze, so electrolysis must have begun early. I have refastened larger hulls built the same way. After 25 years, there isn't enough left of the old fastenings to draw them.

The cabin that Sam Devlin designed for *Surf Scoter* looks quaint, but no boat of 60 years ago had a cabin anything like that. It's a clever trick, like bridal gowns: White is considered traditional, but few brides wore white in the days when most were virgins. The Elco 26 had a truly quaint cabin and layout, which very few owners would want today. Her sheer appeared to break not quite half-way aft, though the actual raised deck continued aft another 2½ feet and was disguised with paint. Under this deck was all the accommodation—four bunks in tiers, and 5 feet 8 inches of headroom on centerline. You ate with the plate on your lap, and looked out through tiny round portlights. You steered from the cockpit, standing up, where a windshield and roof gave some weather protection. I began doodling a Devlin-style cabin and layout for the Elco.

The problem with reproducing such a boat is weight. Engines and

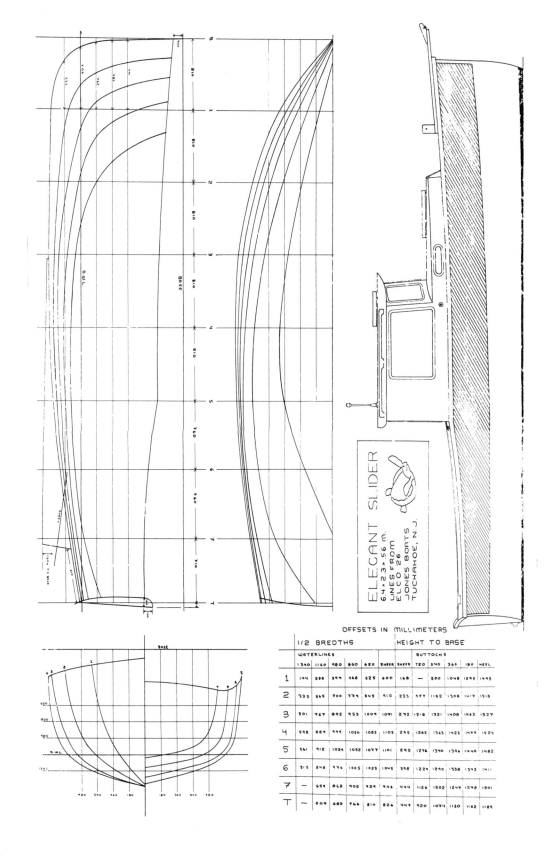

ELEGANT SLIDER
6.4 × 2.3 × .56 m.
LINES FROM
ELCO 26.
JONES BOATS
TUCKAHOE, N.J.

OFFSETS IN MILLIMETERS

1/2 BREDTHS   HEIGHT TO BASE

	WATERLINES						SHEER	BUTTOCKS				KEEL
	1340	1160	980	800	620	SHEER		720	540	360	180	
1	144	288	399	468	525	600	168	—	500	1048	1295	1493
2	335	565	700	779	845	910	233	977	1182	1308	1417	1518
3	501	767	895	953	1009	1071	272	1218	1321	1408	1463	1527
4	598	887	999	1050	1083	1103	295	1285	1363	1425	1477	1524
5	561	912	1024	1052	1077	1101	295	1276	1340	1396	1440	1482
6	315	848	976	1005	1023	1045	398	1229	1290	1338	1373	1411
7	—	659	862	905	929	946	444	1126	1202	1247	1278	1301
T	—	207	682	766	810	826	447	920	1074	1130	1162	1189

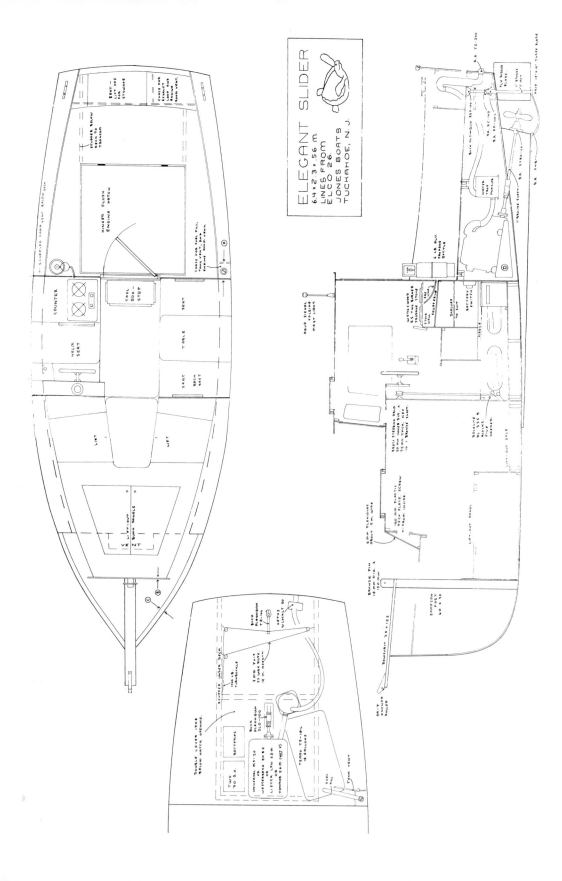

ELEGANT SLIDER
6.4 x 2.3 x .56 m.
LINES FROM
ELCO 26.
JONES BOATS
TUCKAHOE, N.J.

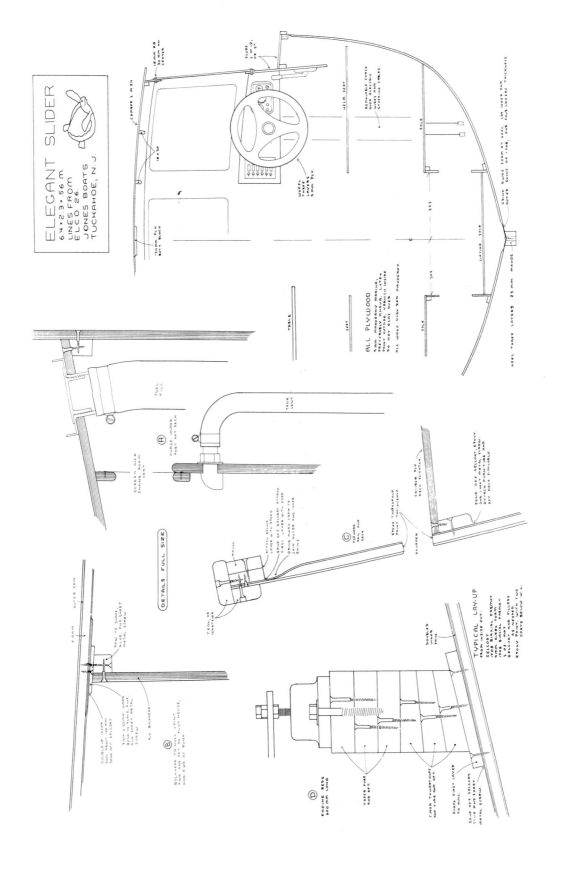

Figure 6–5. Elegant Slider *on the Tuckahoe.*

building materials now weigh about half what they did then. The original boat was meant to sit on her 6500-pound marks with 800 pounds of inside ballast, but with a foam-sandwich or a strip-planked hull, plywood cabin, and modern engine, ballast might need to be 40 percent of total weight, and that made me uneasy. I tried decreasing the deadrise to get lighter displacement, but the result wouldn't have had the same motion

Figure 6–6. Little transom is immersed at speed-length ratio of 1.8.

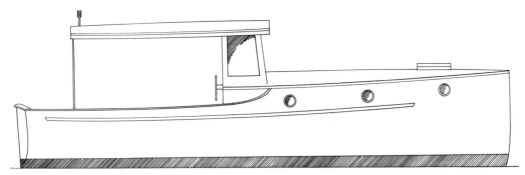

Figure 6–7. Elco 26.

or handling characteristics at all. Another possibility was to scale her down.

The factor chosen was a 20-percent reduction of all dimensions. Sixty-five hundred multiplied by .8 cubed would result in a displacement of 3328 pounds, but as discussed with *Scampi*, the smaller boat would likely weigh more than half the bigger one, because it would have more than half the skin area, etc. The engine might weigh less in the smaller boat, but transmission, prop shaft, and batteries would likely be the same. And payload, two to four people and their booty, was also unlikely to change. So scaling down the Elco lines might result in a boat needing no more ballast, proportionately, than the original.

In addition to scaling down the lines, I reduced the freeboard forward, broke the sheer farther aft, and took the tumblehome out of her. This made building easier, and because she was a very boaty shape already, she didn't need tumblehome to exaggerate it.

Hull material was the next question. For a round-bilge cruiser that would probably be trailered, foam sandwich and strip planking were the only two materials seriously considered. Foam sandwich was chosen because I didn't want to spend all those hours behind the table saw, ripping the strips and eating the dust. A short immersion in polyester resin seemed preferable to a long one in epoxy. Materials for foam-sandwich construction cost more, but strip planking takes much longer to do.

Foam sandwich has a number of advantages in a hull as complex as this. Extra layers of fabric can be added and the hull strengthened wherever needed—under the engine beds or at the bulkhead-to-hull joints. Despite these extra bits, the 220-square-foot hull weighed only 320 pounds when it was brought into the shop. I began wondering if I'd scaled the Elco down enough. The keel was laminated of three layers of 5/4-inch mahogany, which allowed the shaft log to be cut out before laminating, rather than bored out later. The shaft of the Elco 26 came down alongside the keel, not through it, in an economy move the company defended with other and less convincing reasons. The *Elegant Slider* keel was bedded and lagged through the hull.

Usually fiberglass boats have fiberglass-covered engine beds, with encapsulated wood rotting away quietly inside it. In production boats, other fittings are often incorporated into a hull liner, a fiberglass molding more devilishly complicated than the hull molding. It freezes every small detail of the layout, down to where you put your toothbrush, and only suits high-volume production. Otherwise, fiberglass components such as bulkheads and furniture are made separately and glass-taped to the hull.

I wanted as much wood as possible in this fiberglass boat, and I didn't want it glass-taped in. I called the Gougeon Brothers, who pay a knowledgeable man to answer building questions over the phone. He said I could certainly epoxy whatever I wanted to the hull if the gelcoat was sanded back until the glass strands showed, and if a large enough area was covered with the initial layer of wood. Thus, the first piece of wood in each bulkhead was a mahogany veneer $\frac{1}{8}$ inch by 2 inches, bent to the curve of the hull. The bulkhead ground was sawn and glued to the veneer, and the plywood bulkhead itself was then spiled in. Furniture and engine beds were done in the same manner, and it's worked fine.

Attaching the Deck

The Wheelbarrow Garvey had a foam-sandwich deck, which was simply glass-taped to the hull, outside and in. The Tuckahoe Catboats have a wooden sheer consisting of a piece of mahogany the thickness of the foam. It was screwed to the mold, the foam was butted to it, and it was glassed along with the foam, outside and in. This method is recommended by manufacturers of foam core materials, and has given no trouble, but the thought of wood inside fiberglass is disquieting.

On *Elegant Slider*, I put 2-inch duct tape on the outside of the foam, all the way around the sheer. When the outside was glassed, I cut back the top 2 inches of foam, peeled off the tape, sanded, and bonded the inner glass skin to the outer. The sheer was then solid fiberglass. A sheer clamp was laminated in place, screwed through from outside, and the guard rail eventually covered the screw heads. This is the strongest sheer for foam-sandwich boats. *Elegant Slider* also has solid glass in way of her keel, stuffing box, and rudder port.

The Elco 26 was originally marketed with a 4-cylinder Gray engine, 165 cubic inches, producing 15 to 18 horsepower at about 1000 rpm. Top speed was guaranteed to be 9 mph. After a couple of years, the company switched to a 6-cylinder Gray, 27 to 30 horsepower, and top speed rose one mph. For *Slider*, I chose a Universal diesel, 3 cylinders and 36 cubic inches, producing 18 horsepower at 3600 rpm. For the half-size boat, this is a superabundance of power, and the reduction gear means that the prop doesn't turn much faster than the Elco's.

To run his boat, the Elco skipper stood smack on top of his engine. This may not have been too bad with an engine of 4-to-1 compression ratio, but you may have noticed that most older powerboats that are repowered with diesels soon sprout fly bridges. This is not a back-to-nature movement, but a need to get away from the noise and vibration of the diesel. *Slider's* engine *had* to be away from the crew, and this controlled the layout. Its compact size allowed it to go under a flush aft deck, although this required a 10-degree shaft angle, which is more than desirable because any shaft angle costs efficiency. A V-drive would have allowed a lower shaft angle, but with engine weight farther aft and engine bulk higher up, and a flush deck would not have been possible.

The Universal engine is truly a universal effort. The block is made in Japan by Kubota Tractors, the Hurth transmission is German, and many of the small parts are made by Universal or its subcontractors in the United States. The heart of it, however, is the very sophisticated and high-revving Japanese component, and for a diesel it is both smooth and quiet. Two mufflers and rubber engine mounts don't hurt. The bulkhead between engine and pilothouse is insulated with PVC foam, and though there are more elaborate soundproofing systems on the market (they use multiple layers of foam and thin sheets of lead), it is possible to hear a conversational voice across the pilothouse when *Slider* is cruising at 2400 rpm.

The 13-inch propeller diameter was taken off the nomograph in *Skene's*, and confirmed by Universal. The 10-inch pitch was intended to give 8 knots cruise at 2400 rpm, which it about does, and the engine will rev to 3100 rpm, which might be 10 knots. A 9-inch pitch was another possibility, and might give another half knot of top speed, but would require another 300 rpm to achieve 8 knots. I chose the cruising prop. People go through all kinds of conniptions with inboard propellers, probably because they have unrealistic ideas about the speeds their boats should go. Even fast boats are slow, compared with slow automobiles.

Fitting engines, prop shafts, steering cables, and much other mechanical gear is easier if the lofting is kept on the wall after the boat is built. I never take one down until the job is out the door. This kind of mechanical work is great fun, even for a carpenter. Laying it out, making it work, and making it neat call for constant inventiveness. It's different work altogether from finding and replacing defective parts in a boat or automobile that someone else laid out. I find electrical work less fun, because a part of my mind still refuses to understand electricity, and the fittings are so tiny and so close together.

Layout

Elegant Slider is bigger than *Scampi* (waterline is as long as *Surf Scoter's*), so there is more room on deck as well as below. I left the aft deck uncluttered except for a seat across the stern with a locker underneath for fenders and docking lines. Powerboat people don't have to move around as suddenly as sailboat people, and don't mind stepping over a few deck chairs and a cooler. Below, the bunk is a V-berth, rather than a double. It's less friendly, but easier to get in and out of, and there's room in the center for a portable toilet, perhaps with a curtain between it and the pilothouse for the illusion of privacy. To my nose, the chemicals in these toilets

Figure 6–8. Engine room. Note emergency tiller under deck to left of muffler.

stink worse than nature. On a recent sail up the Hudson, Carol and I noted that all boats used chemical heads at dockside, and as there were few pump-out stations, they later dumped them in mid-river. Why is dumping chemicals and sewage an improvement on dumping sewage alone? Rules intended to help the environment have gone awry. In Canada, uniform rules about toilets and pump-outs were thought out and put in place. The system may have faults, but it does work. Americans are less capable of this kind of organization. When a shore toilet isn't available, Carol and I use a bucket head, but we aren't proud of it.

Elegant Slider's pilothouse is 4 feet 4 inches long, and wider than *Scampi*'s. A camper stove does fit comfortably behind the helmsman, though a marine stove would have been too big. As on all the boats I've built with propane cooking, the stove is in a self-draining, box-like well. The propane tube runs out the drain, and the bottle is on deck or in a self-draining locker. If the system leaks anywhere, or if a burner is not turned off completely, the fumes go overboard and don't collect in the bilge.

To port is a dinette for two. *Skene*'s says that for two 6-footers, a dinette should be 5 feet 7 inches from seatback to seatback. *Slider* gets away with 17 inches less because the seats and sole are more than 18 inches wide, so the first person in swings his legs toward the topsides and the second person keeps his legs near the alley. If the two people aren't friendly, the second person had better get out first. Anyway, on small boats you can't expect all furniture to be of household dimensions, and you use the space there is. Seats in this boat are the standard 17 inches off the sole, but I've sat in many boats with seats 10 inches high, and been glad they weren't 5 inches. Over the forward dinette seat, a plywood flap hinges down to make a chart table twice the size of the eating table.

It would be possible for four or five people to eat dinner belowdecks in *Slider*, but some would have plates on their laps. Eight or ten people would not be impossible for cocktails, milling around from cockpit to pilothouse to forward well deck, spilling drinks on each other and grinding peanuts into the sole. Ernest K. Gann, describing a 35-footer that he commissioned from Jay Benford, said proudly that she "drinks six, eats four, and sleeps two." Tell us, Jay, where do you find such reasonable clients?

I considered a teak deck, but this wood is hateful to work with, and I find it hateful to maintain. Originally, teak was used for hulls because teredos prefer to chew on other woods. So do I. A snootful of teak dust from the table saw will set my sinuses dancing like no other wood I know. Probably it's the silica in the wood, which is also what dulls the saw blade so fast.

Lately, teak has replaced fiberglass as the most glamorized and misrepresented material in boats. Some genius has convinced most of the boating public that teak, like fiberglass, is an impervious material (well, even more than fiberglass, because it doesn't blister). It needs no finish and no maintenance. Therefore, the manufacturers use it for guards, handrails, steps, and other fiddly bits. It would cost them twice as much to make the same bits of mahogany and give them three coats of varnish than to make them of teak and leave them raw.

Teak is highly rot-resistant, but it does grey and weather away in sunlight like all other woods. This happens to it about as fast as to other woods the same density. Like all woods, it can be protected from sunlight for a year with varnish, or for five years with paint. There are special oils made just for teak, and some of them don't last a week.

In token submission to fashion, *Slider* does have teak guards and handrails, but I've oiled them and will keep oiling them until the boat is sold. Decks are Treadmaster, a British composite that looks like flecks of cork set in rubber. It comes in flexible sheets, 3 feet by 4 feet, and is glued down with epoxy. For years it has been the favored decking on Royal National Lifeboats. It has far better skid-resistance than teak, requires no maintenance, and comes in half a dozen colors, one of which reflects the sun's light and heat instead of absorbing it like teak. I have high hopes for this stuff.

When launched, *Slider* needed 400 pounds of ballast forward to trim out for the engine weight. We used lead wheel weights set in mortar, and poured this around iron sash weights. I don't like moveable ballast, which is seldom moved except by the forces of nature, and always at times when it had better stay put. Moveable ballast collects dirt, and makes more dirt by corroding. If *Slider*'s ballast ever needs to be removed, very little work with a brick chisel or electric hammer should make it easy enough to handle.

Completed and in the water, she doesn't seem much like a fiberglass boat. Most of the surfaces you see and touch are wood, painted on deck and varnished below. In the bunk cabin, the topsides offer quite an expanse of plastic, but at least it is textured by the cloth, not smooth and shiny and cold. Looking up, you see the varnished beams and decking. In the pilothouse, almost no fiberglass at all is visible. The wheel is copied from one we saw in a Dutch boat, cut out of plywood $1\,1/8$ inch thick and 21 plies, made by laminating up three pieces of $3/8$-inch ply. The end grain is gorgeous.

Moving out from the dock, the big prop and rudder give great control. Helmsmen who have spent time with single screws can do tricks with a boat like this: step her sideways with wheel and gearshift, make her turn in very small circles. Revving her up, the hull does not squat, but the bow cuts through its own wave. The resulting wave rises steeply up the bow, but has little volume of water in it. The stern has considerable deadrise, and doesn't offer much of a lifting surface, but this doesn't keep her from exceeding hull speed. Eight knots is a speed-length ratio of 1.79, and 10 knots is 2.24. This hull may be more truly a semi-displacement hull than *Puxe*.

We have not tried *Slider* in a big sea. Downwind in a Force 6 in Great

Figure 6–9. Elegant Slider's *console.*

Egg Bay, *Puxe* would be yawing a bit, though not dangerously. The helms-man would need to be alert. *Slider* runs perfectly straight in those condi-tions, and the helm can be left alone until the boat gradually wanders off course, as she would in a flat calm. She's a better sea boat than *Puxe*, due to her greater weight and depth of hull and deadrise aft. She can roll pretty good to a beam sea, with no chines to dampen her, but waves from bow or stern don't phase her, and she's never shipped a drop in anything we've seen.

Dockside, she's not quite as cute as *Scampi*, but she still draws plenty of attention. All other things being equal, smaller is always cuter. The round hull also looks more purposeful and less toy-like than the square one. People are more likely to admire her respectfully, and less likely to fall in love with her at first sight. The double guard rail on *Scampi* adds to her charm, but this would be too much work in a foam-sandwich hull, with all that grinding back of the foam, and the resultant hull would be less stiff. The double sheer on *Slider* is only paint. *Scampi*'s salient bow-post is a wooden-boat touch, and it is not imitated in fiberglass easily or to any purpose. John Groot prefers the looks of *Slider*, but she is enough bigger and heavier to be a lot more work to trailer, and she is more expen-sive.

Owning her, I'd take her anywhere within range of a weather forecast—New Jersey to Block Island; Florida to the Bahamas. The big dividing line between reasonably seaworthy boats like *Slider* and real ocean cruisers is the ability to go farther than the predictions of the weather forecasters — New Jersey to Bermuda, for example. For that trip in a powerboat, I'd recommend a lifeboat and a strong constitution. *Slider* lacks the tankage for it in any case. If she burns 2/3 gph at 8 knots, her 18 gallons would last 27 hours, or 216 miles. At 5 or 6 knots, she might get 300 miles out of a tank.

We have not run her far enough to pinpoint fuel consumption. It's safe to say that *Elegant Slider* uses one-quarter of the fuel that *Scampi* does to push a somewhat longer but substantially heavier boat at the same speed. Diesel versus gasoline 2-stroke is part of the answer, but another part is the lower-resistance hull shape, designed all those years ago when people were more conservative. They turned off lights in rooms they weren't using, and made other niggling economies that we can't bother with today. They painted turtle shells, too, of course, and on the whole they weren't any kinder to the environment than we are. But we can still learn a few things from them.

7. Multihull Sailboats

Late in July 1975, we were hanging around St. George, Bermuda, waiting for the Navy weather service to predict the future of a tropical depression over Cuba. It was hot in St. George, but we had plenty of good company and good English beer on draft at the White Horse Tavern, only steps away from the dock where we were tied. We were nervous because the season had already brought a tropical storm and yachting fatalities: The Horstman trimaran *Meridian* had recently been found, upside down, with her diabetic skipper dead from losing his insulin in the capsize. *Gloria Mundi*, a production fiberglass monohull with four college boys aboard, was now a month overdue from Newport. Tom Bolger, the father of her skipper, had visited us aboard *Two Rabbits* several times, asking about the weather we'd seen on the way down, and we had given him what encouragement we could. Staying in Bermuda at that time of year increased the likelihood of another storm or even a hurricane on the way home, and we wanted to be off.

Waiting across the dock from us was *Sabrina*, an Atalanta. This class was designed by Uffa Fox to be parachuted to downed fliers in World War II. Twenty-five feet overall and about the same on the waterline, she had no corners and no flat surfaces, the sheer rounded over into the deck, and the cabin was a low blister. She was hot-molded plywood, with retractable bilge keels and an interior full of meaty bulkheads. *Sabrina* was sailed by three peppy young Englishmen who had brought her here via the Azores, and hoped to sell her in America. They visited us on *Rabbits*, and signed the log: Mr. Eric Stallard; Navigator Robert Clarke; Crewman Adrian Cargill. "If you run into a hurricane, you'll have the boat to take it," I said.

"And the crew to take it, too," said Adrian. They were short of cash, but not of pluck, and they left bound for Newport a day ahead of us.

Robert Benson on *Banjo* we hardly met. He came in soon after *Sabrina* left, nine days out from New York with a crew of three. He had run into calms, and had motored much of the way, he said. *Banjo* was a year-old, solidly built Dutch fiberglass sloop 31 feet long.

Compared to these boats, *Two Rabbits* was small and frail. A 23-foot Wharram catamaran, she was built of 1/4-inch plywood and weighed 1200 pounds. We kept calling the Navy for updates, and finally at 1400 on Tuesday they announced that the depression was headed toward Florida,

Figure 7–1. Two Rabbits in Hamilton Harbor, Bermuda.

TWO RABBITS
23' × 12' × 13"

MAIN 114 SQ. FT.
JIB 59 " "
TOP SAIL 27 " "
SQUARE SAIL 180 " "

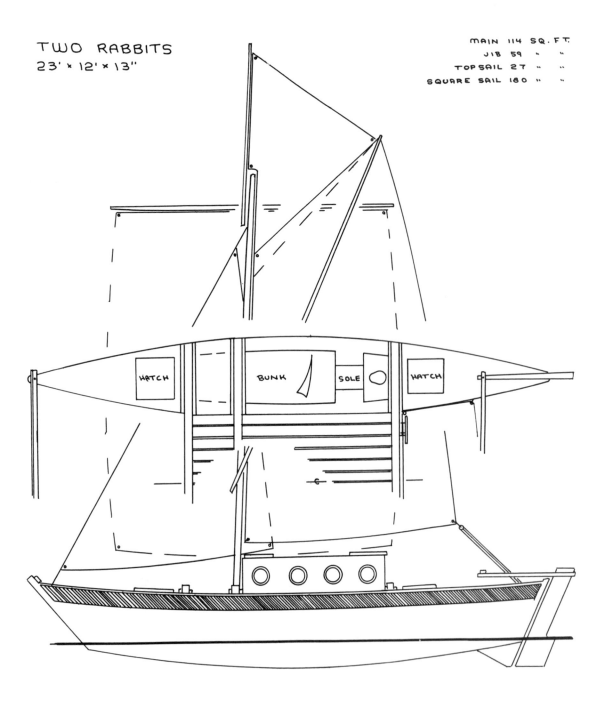

HATCH BUNK SOLE HATCH

with wind speeds of 15 to 18 knots. It wouldn't be a hurricane. We cleared customs, and by 1600 were out Town Cut. The wind was southerly and light. Rounding Northeast Breaker Buoy, we broke out the squaresail and were able to fly it for 50 consecutive hours.

Banjo departed St. George next morning. They had meant to stay longer, but were adhering to a schedule. The weather forecast still sounded good.

Our third day out was gorgeous. "A day of unusual clarity," the words often used to describe a day before a hurricane. We hadn't read such a description at the time, and we were enjoying ourselves. Storm petrels were visible almost on the horizon, and it seemed that only the curve of the earth kept us from seeing New Jersey ahead and Bermuda astern. Colors were brilliant, and the rows of gleaming Sargasso weed marched away from us with never-ending interest. Carol hustled around the deck, throwing back while they still lived the flying fish that came aboard. But in the evening the squaresail had to come down. The wind was rising, though its direction was holding steady. "Just at dinnertime," I wrote in the log. "Managed to hold it 'til after, eating on hatch with one hand, steering with other. Things fall off shelves. Here we go again."

For the next two days we made wet but satisfying progress. Our noon fix on July 26 showed 360 miles logged in less than four days. Though we didn't realize it, we were overtaking *Sabrina* and *Banjo* was overtaking us. We were all sailing the same course, and were remarkably close to the same spot, 35N 70W, when Benson turned on his radio for the July 26 noon time tick and learned that tropical storm Blanche was 100 miles south of us, and heading our way.

Our two small radios were the Sunday-in-the-park variety, so we did not learn about Blanche until 36 hours later, when she was off Nova Scotia and blowing at hurricane strength. Had we known, we might not have behaved much differently. *Banjo* knew, and her best survival tactics put her in sight of us that evening, as Blanche closed in.

Two months later, when I learned in a letter from Robert Benson that he had sighted us at the beginning of the storm, my hair stood on end. We were under a 21-square foot storm jib then, running off on course at good speed. We were below in the windward hull and didn't see him. He dared not come too close because the seas were enormous and the wind blowing just under hurricane strength.

Half an hour later, he was forced to hand his own storm jib and lie ahull. His masthead rig was vibrating so much that he feared he might lose it. In this posture, at about 1900, *Banjo* fell off the crest of an enormous sea, bending her skeg and rudder and opening a 5-foot crack in her hull.

At the time, *Sabrina* was also lying ahull, but she took it better, perhaps because she was designed for it. Perhaps she didn't encounter a sea like the one that broke *Banjo*. The ocean is not a test tank, and no two waves are alike. *Sabrina* was probably farther west than *Banjo* or *Rabbits*, which put her farther from the eye of the storm, which finally passed 30 to 40 miles southeast of us. But *Sabrina* was nearest the Gulf Stream, and may have encountered the worst conditions there.

Sabrina endured, but it was no pleasure. "You sure had it easier than us," Eric wrote me later. He and his crew were flung back and forth across the cabin, and most of their gear broke loose and followed them. "It was sheer hell," says Eric. "We were lucky to survive."

Two Rabbits surfed off until 2300 under storm jib with steering lashed. We jibed her when the wind backed to east, because we were making progress and wanted to stay on course. We kept telling each other that this couldn't be a hurricane, and wouldn't get any worse. It was just that the barometer, which had always been reliable before, had picked this moment to go haywire. No matter how we beat it on the bulkhead, it continued to drop .05 inch, and sometimes .10 inch, an hour. We tried to read *Eldridge* and the *Sailing Directions*, but wind coming in from somewhere kept blowing the candle out. And we were so scared that we couldn't grasp what the books were telling us.

Toward midnight, we began lifting a hull when we came up on crests. We were on the starboard jibe, with wind and waves on the quarter and the old waves astern. Jibing would have put our weight to windward, but it would have put the old waves abeam. We discussed it a long time, full of the lethargy that comes with the fear of death. It seemed that our only choice was to get some weight into the other hull.

My crawl across the 5-foot-wide duckboard deck was surely our worst moment. We had been pooped once, and water had come in through the hatch seams (perfectly tight on all other occasions) with the strength of a fire hose. If we had been pooped again while I was crossing, a harness would only have broken my back, and the time spent attaching it would have tripled the time the hatches were open. We said good-bye to each other. I waited until we slowed down in a trough, popped the hatch, and went.

I have no idea of the wind speed, but the noise in the five little wires of our 18-foot mast was like a freight train. Spindrift stung like flung sand. I couldn't see the waves, but Benson guesses that his nemesis (undoubtedly a freak) was a 50-footer.

My weight in the starboard hull made a substantial difference. We continued to run off, staying ahead of most seas. After midnight, I sensed what Arthur Piver described in *Trans-Atlantic Trimaran*. We would surf down a sea and stop, as if brakes had been applied. We were burying our bows in the sea ahead, and were in danger of pitchpoling. I was able to stand up in the forward hatch, reach the halyard, and lower some of the sail.

Facing downwind I could see pretty well in the almost continuous lightning. The water was so aerated that the boat had hardly any freeboard. The air was so full of water—rain and spume—that the distinction between air and water had almost ceased to exist, except that we could survive in one and not the other. Certainly there was no horizon between the two.

The taut sheets kept enough sail standing to hold the bows downwind and keep us moving. In that condition, we endured the strongest winds, which blew from the northeast after midnight. Carol and I missed each other's company then, but we both knew that the storm had come around us 135 degrees, and could only diminish. Half our gear was on the floor, but the other half was staying on the shelves. We both lay down and slept until dawn.

NOAA saw Blanche as the most typical of hurricanes, and for years they used it on the backs of pilot charts, to illustrate what hurricanes were most likely to do. All of us who were in it appreciated most the speed of its movement northeast—the worst of it was past in about eight hours. *Sabrina's* crew was demoralized. Their boat was whole, but they were too bruised and sore to work her. And as their sextant was bent, they couldn't even tell where they were. Some days later, under motor, they made Oregon Inlet.

Benson and his crew, with the energy of desperation, kept *Banjo* afloat until dawn. They pounded rags into the split hull, manned every pump and bucket, ran the diesel with the water intake sucking from the bilge. They had good radios, and were in touch with the Coast Guard long before daylight. In the morning, a cruise ship successfully picked them all up, and from her deck they watched their boat go down. "We often wondered what happened to you," Benson wrote me later. "Cats look so frail."

Soon after sunrise, Carol and I stood up in our respective hatches and greeted each other shyly, almost like strangers. We had some gear to clear up below, but soon enough we had sail on; and for the rest of the trip home, we had no worse problem than light air. It was some days, however, before we understood fully what had happened to us.

Polynesian Catamarans

James Wharram, the designer of *Two Rabbits* and a whole range of "Polynesian catamarans," has no training in his trade, and even now, 35 years

Figure 7–2. James Wharram aboard his 53-foot Polycat Tehini *in 1973.*

into it, he uses concepts and vocabulary to describe it that are not heard elsewhere. Wharram was a rebel from a very early time. His feelings and ideas put him at odds with society, which has sometimes treated him with contempt, and has often received contempt from him. Today, at 60, his manner is more conciliatory than it once was, but the rebelliousness and contempt are still not far below the surface, and can flash out.

Wharram grew up dreaming of boats and ocean voyages. Early on, he tried to fit out a hulk of a monohull for the job, but a gale destroyed it in harbor. He then built a carvel-planked 23-foot catamaran. "Two coffins," said his friends. With two German girls and a dog aboard, he did succeed in making a tradewind crossing in her, but she was appallingly wet and slow and teredos ate her up. Ruth kept her afloat by stuffing the teredo holes with chewing gum. In Trinidad, Jim built *Rongo*, a plywood 40-footer that became the archetype of his whole range. She was based, he said, on four Polynesian principles: narrow beam-length–ratio hulls, V-shaped sections, flexibly mounted beams, and no permanent deck cabin between the hulls. *Rongo* probably also owed something to an Uffa Fox catamaran. This design was commissioned by Bell Woodworking, and a prototype was built while Jim was working for Bell, cutting out parts for kit boats.

They sailed *Rongo* home to England, and there they promoted a huge cache of food and supplies for a round-the-world voyage. Jim and Ruth sailed her out to the Caribbean again, but Jim was mourning for the other German girl who had died, and he couldn't get over being seasick whenever the weather was rough. They headed back to England again, with enough food in the bilges to last them several years at dockside.

Wharram went into the business of selling catamaran plans, and he has been good at it. For many years, he probably sold half of the plans that were sold by all the world's designers. On our trip around the Atlantic in 1979–80, Carol and I seldom entered a foreign port without finding several other Polycats already there. Jim has a flare for publicizing himself and for expressing his ideas so that newcomers to sailing can understand them. From the start, his literature has emphasized that a boat is not just a "piece of sporting equipment," as Dennis Connor has called it, but a means of self-expression and of escaping from industrial civilization. *Two Girls, Two Catamarans*, his book about his first two trans-Atlantics, sounded themes that he has often repeated since. His most recent design book says, "To build and sail one's own boat gives an inner certainty, strength and apartness to live one's own lifestyle."

Jim is also capable of standing up in a meeting of the Amateur Yacht Research Society, chaired by the Duke of Edinburgh, and saying that the ideal boatbuilding material, because it is so cheap and available and plastic, must be bullshit.

Today, it is disconcerting to meet the notorious and flamboyant James Wharram. You aren't expecting such a simple, friendly, and uncontentious person. Unlike so many Englishmen, there is no *side* to what he says: no innuendos, no hidden barbs, no condescension. He is a genuinely nice man, and loves to talk about boats. Terms like prismatic coefficient, wetted surface, or foils he never uses, and it's hard to be sure what he knows and doesn't. Some things he mis-equates. Sail area cannot be simply related to displacement, for example, except in boats of nearly the same size, because sail is an area and displacement is a volume. The square root of sail areas must be related to the cube root of volume, as it is in the Bruce number, or in the more complex formula for sail area divided by displacement to the $2/3$ power, which is now going out of fashion.

Figure 7–3. Wharram's building method.

Building Method

Rongo was built conventionally, upside down on a strongback, but Jim soon evolved another building method. He presupposed that the builder would have no building site worth mentioning, very few tools, and even less experience or manual skills. The heart of his method was a backbone, a kind of plywood keelson butt-blocked together, deep enough to be rigid in one plane over the length of the boat. The V-shaped frames, the keel, and the end posts were all attached to it, and the whole thing was set up anywhere and aligned with a string and shims under the various legs. Stringers were fastened to the surface of posts and frames, instead of being notched in, and the gaps between them were filled with "packing pieces." All lumber was stock lumberyard sizes, so a 53-footer could be built without a table saw, if desired.

We built *Two Rabbits* to Wharram's method, although we did use power tools. Once the stringers were on and the form was stabilized, we cut out the backbones, which would have cluttered the storage areas, and wondered why we had wasted good plywood on them in the first place.

We didn't have much knowledge of boatbuilding in those days, but we were confident our enthusiasm would make up for it.

Hulls were built one at a time. They were 23 feet by 3½ feet, and the site was the living room, dining room, and kitchen of the row house in which we were living. Space available was 25 feet by 14 feet, so furniture was moved to one side or the other and work progressed. We used fir marine ply throughout, and sheathed everything with nylon cloth and polyester resin. The fumes penetrated everywhere, but lodged most permanently in dairy products. The gasket of the refrigerator door didn't protect butter or cheese, and often they had to be thrown out.

The outboard side of the cabins tumbled home to be parallel with the inboard side of the hulls. This was handy for rigging shrouds, but essential for getting the hulls out the window and into the street. Even so, the window frame had to come out, and the very first hull to be moved caught on the loose stucco of the facade and pulled away many square yards of it.

The biggest change we made to Wharram's *Hinemoa* design—approved by Jim when we visited him the summer before—was to eliminate the bulwarks and raise the hull freeboard to the height the bulwarks would have been. These Polycat bulwarks are intended to hide the connecting beams, and rectify any unfairness in the sheerline. Without them, and with longer cabins than shown on the plans, we nearly doubled the interior volume of our boat. Still, Carol and I slept on a double bunk 28 inches wide. My great aunt, then over 90 and upholstered like a sofa,

Figure 7–4. One Rabbit *emerging from living room.*

came aboard once for cocktails and looked down into the bunk. "Do the two of you sleep there?"

"Yes, we do, Aunt Mary."

"Well, good for you, dearie."

In Hurricane Blanche, I'm sure the extra buoyancy of the hulls made the difference between pitchpoling and not, between being smothered by those huge airy seas and floating on them. To be fair, Wharram didn't intend the *Hinemoa* for ocean passages, was horrified when we made ours, and cautioned others not to. Though no longer overall than his own first catamaran, *Rabbits* was several feet shorter on the water and perhaps a quarter the displacement.

The raised sheer and larger cabins made *Rabbits* heavier than a stock *Hinemoa*, but we watched every ounce that went into her, both in construction and in loading. She never had a dinghy in the five seasons we cruised her, and when we couldn't tie her to a dock, we swam ashore. She never had a motor, but I rowed her at about 2 knots, sitting in a forward locker with a single oar between the hulls and giving plenty of instruction to Carol at the wheel. We had tried the long tillers Wharram showed, but found them disconcertingly whippy and went to two plywood wheels, one beside each companionway. Quarter-inch nylon rope connected them to short tillers, and there was enough spring in the rope to keep the system tight in wet weather or dry. It never gave trouble.

Rabbits never had Ackerman linkage, though we put it on *Vireo* later. Any catamaran can benefit from it. The principle is used in automobile steering systems: In a turn, the inside wheel (or rudder) should turn more sharply than the outside one, because it is describing a smaller radius. To do so, the connecting bar between the tillers needs to be shorter than the distance between the rudders. This can be achieved with curved tillers or, more simply, with wooden blocks glued to the insides of straight tillers. The choice might depend on whether the skipper wanted tiller handles amidships in the hulls, or more inboard. They say that in a catamaran, the greater the Ackerman linkage, the better she turns, up to the point where one tiller is in danger of overcentering.

To save weight and clutter on *Two Rabbits*, we bathed in the head bucket. We skipped a dishpan, and washed each dish overboard (it wasn't far to reach). We had no pillow on the bunk, but made do with a kapok flotation cushion which we also sat on when steering, and which was always wringing wet. We never went as far in trying to save weight as the two Englishmen we later met in Barbados on a larger Polycat. They brought only one book with them from Europe, an 800-page historical novel which they read in turns, each one starting it again as soon as the other had finished.

Briefly, we did have a coal stove on *Rabbits*, which weighed about 20 pounds. We launched in April the first season, and found the warm cabin agreeable, though Carol did singe a whole sleeve off a jacket, reaching around the stove for more fuel. By June, the heat of making even a cup of coffee was impossible, and we'd had enough of downdrafts from the sails backwinding the stove and filling the cabin with smoke and ashes, some of them still glowing. We removed the stove, and thereafter cooked on Sterno.

Sail Plan

Hinemoa plans showed a boomless spritsail and jib with a total area of only 173 square feet. We took it to Gil Webster, and he agreed that it was small. He cunningly added a few inches wherever he could, and larger roaches, too. On launching day, with two people and no gear aboard, we were pleased with the performance. As *Rabbits* acquired even her few small bits of gear, she settled noticeably, and we had to raise the water-line. Wharram's drawings show the waterline of a nearly empty boat, and hulls with 18-inch waterline beams don't need much pushing to sink

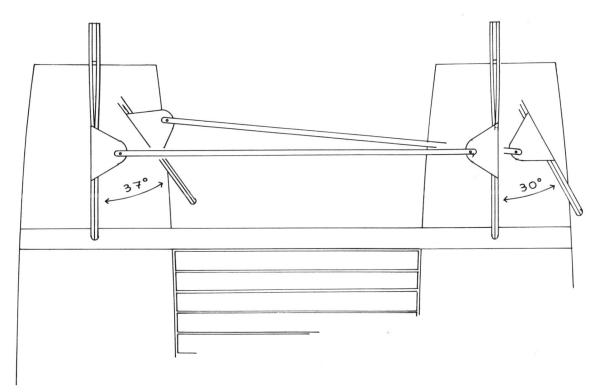

Figure 7–5. Ackerman linkage.

Figure 7–6. Rabbits *immediately after launching. Note Charley Noble for coal stove.*

deeper. The *Hinemoa* is supposed to have a payload of 1000 pounds (though I'm sure we never put that much in ours) and to draw 13 inches. Fully loaded, she'd have to draw at least 17 inches. Her wetted surface would rise extraordinarily, and her length-to-beam ratio would shrink.

From the first day, we had a squaresail bigger than the working rig (not shown on Wharram's plans). Gil made it of 1.5-ounce nylon. This was always a wonderful sail. The yard hoisted to the hounds, so it could be braced around any amount without fouling the rigging. The wide beam of a catamaran gives a perfect base for braces and sheets, and this sail had nothing in common with the toy squaresails that you see on L.F. Herreshoff and John Hanna designs. Ghosting on a reach, it would fill and draw when the main wouldn't, although the apparent wind it generated soon had the main filling and drawing, too. Farther off the wind, it required no more attention than a mainsail. We braced and sheeted it, and left it to do its work. It never collapsed like a spinnaker, or wrapped itself around the forestay. You could make it foil until the leading edge winked at you, but if you overdid it, the penalty was very much less.

With two of us aboard, and our irreducible minimum of cruising gear,

Rabbits had about 100 square feet of wetted surface. Squaresail and mainsail gave at least 300 square feet and drove her just fine, even in very light air. Close to the wind, when the squaresail couldn't be set, light-air performance was logy, and in her second season we treated *Rabbits* to a topsail. It was set on an unstayed topmast, and the luff was stiffened with a couple of plastic jib hanks to the halyard fall. The sheet was permanently rove through a tiny block on the sprithead, and the sail could be set and struck from deck. Topsails should be more efficient above spritsails than gaff sails, with no halyards to foul them after tacking.

The topsail improved the looks of the boat, and it kept us busy in light weather, but I don't think it added one mile to our daily average. The windage of the topmast when the sail wasn't set cancelled whatever drive it gave when it was.

Gil's working sails were of generous proportions, and they were well cut, but they were too light a fabric for the hard use we gave them. By the time we were home from Bermuda, they were rags. We ordered a new set from Wharram's sailmaker in England, but they were not an inch bigger than what the sailplan showed, and renewed our appreciation for what Gil had done. The last year we had *Rabbits*, we gave her a new mast and a Bermudan mainsail with a sprit boom. Area, including roach, was exactly the same as the spritsail. Daily averages increased from 65 miles a day (about what a Tahiti ketch does) to 70 miles. I think we could have had the same improvement by adding a boom to the spritsail, and controlling sail shape.

Though *Rabbits'* daily averages were no better than other good boats 19 feet on the water, in the right conditions she could roar away from much bigger boats, much to their scornful dismay. If the load could be kept to 500 pounds (and it often was), her Bruce number was 1.10 and her narrow hulls made little wake. At high speed, where wave-making matters most, she was a low-resistance boat. Hull speed never meant a thing to her, and she went through that barrier without our even noticing. However, enough wind to drive her that fast creates such a seaway offshore that we usually chose to reef her. She was not a dry boat.

To windward, *Rabbits* was always a disappointment, and the lighter the wind, the more disappointing she was. Like Hobie Cats, Polycats depend on hull shape to prevent leeway. Hobies work fairly well, because they have little hull windage and an immense, efficient rig. With their greater freeboard and small, low rigs, Polycats don't capsize, but they don't make much to weather either. Wharram argues that catamarans (like monohulls) can trip over a keel or centerboard and capsize, and that is true. I wouldn't go offshore in any boat with a fixed keel. If it doesn't split the hull like *Banjo*'s did, sooner or later it will trip the boat. Cata-

Figure 7–7. Rabbits' Bermudan rig with sprit boom.

marans turn turtle, and monos smash their cabin sides, fill, and sink. But a board that can be raised is another matter.

Hard on the wind, all sailboats make leeway. If the leeway preventer doesn't have different pressure on its two sides, it isn't doing work. If a weatherly boat makes good a course 45 degrees off the wind, a boat making 5 degrees more leeway only sails 20 percent farther to make good the same windward course. However, the leeway also makes extra drag and slows the speed through the water, so the boat with poor leeway prevention may take 50 percent longer to cover the course.

Spring and fall, we kept *Two Rabbits* in Philadelphia and week-ended her often on the Delaware. We enjoyed playing the river tides, which helped compensate for her unweatherliness. Summers, we moored her in Cape May, and in addition to the Bermuda trip, we sailed her to New England three times and south to the various inlets of the Cape Charles coast. All of these voyages were long for such a tiny boat, although twice we made the 200 miles to New England in two days. Nevertheless, we counted on needing four days to get back.

In her third season we spent 39 hours beating down from Barnegat to Cape May, averaging less than 2 knots made good. Though always headed, we had a variety of slants and wind strengths, and we worked ceaselessly at making her go, reefing and unreefing, setting and striking the topsail, playing for lifts from the currents around the inlets. From the pilot charts and from experience, we knew we were getting a steady half knot lift all down the coast from a counter-current to the Gulf Stream. We were not making good a knot and a half by our own exertion. Although we had remained devoted to *Rabbits* up to that point (she had saved our lives as we saw it), that was when we began talking about another boat.

Two years later, we sold her to a man who expected her to patch his failing marriage. For the next few seasons, a third owner kept her moored in Cape May, but seldom used her. Carol and I moved away, but happened back once and saw her on the land after she had been sold to a fourth owner who was preparing to truck her to a lake far inland. She was in poor condition, both from neglect and from rot beginning its subtle work under the sheathing. She filled us with nostalgia just the same. "I can't imagine how we went places in her," said Carol. "Look at how tiny she is!" I was surprised—and grateful to Carol—that our own marriage had survived so many arduous voyages in *Two Rabbits*.

Vireo

Before I started building *Vireo*, I had another 27-foot catamaran lofted on the floor of our summer cottage in Cape May. It had chines, and a centerboard to prevent leeway. It looked good, but I lacked confidence in what I had learned in books. Suppose it didn't work? James Wharram's designs did work, and their shortcomings would surely be mitigated in a larger model. Wharram's rhetoric was still powerful in our minds, and we knew that his boats crossed oceans. We were enthusiastic members of the Polynesian Catamaran Association, and had made many good friends through it. Among them was Jim's wife, Ruth, who had crewed with me on a 40-footer the summer before, racing from England to the Azores.

That race had shown me what Polycats could do, at least when the wind was free. Three days out from Plymouth, we sailed through the

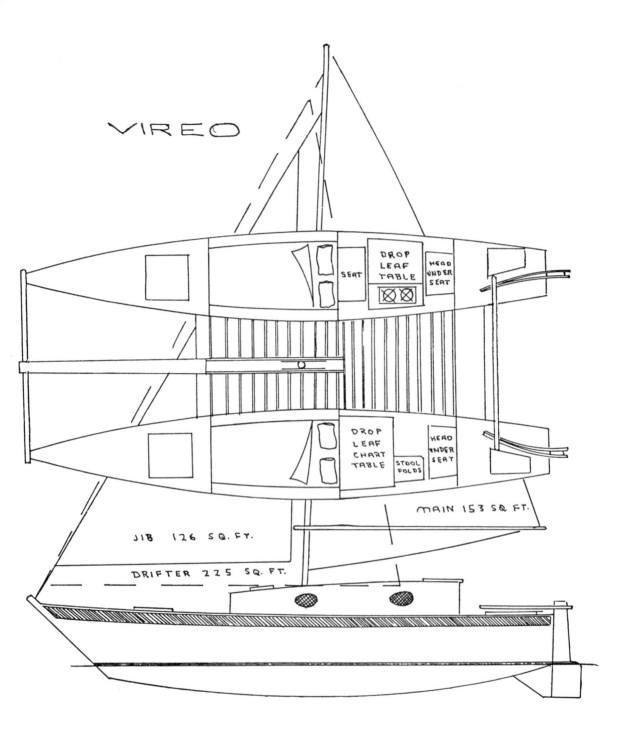

Figure 7–8. Vireo *sailing in Horta, Azores.*

Figure 7–9. Vireo *under plain sail and fender board, Great Egg Bay.*

night with a diminishing beam wind and with running lights always in sight. Dawn found us becalmed, and within a mile of two of our competitors: the final *Gypsy Moth*, 57 feet long and skippered by Giles Chichester; and the original *Three Legs of Mann*, a 39-foot Kelsall trimaran which, the year before, had set the record for the longest day's run ever made by a singlehander. The three crews stared at each other most of the morning. Then the wind came in from ahead and stayed that way for a week. With our tired working sails and lack of leeway prevention, we didn't finish well.

Nevertheless, Carol and I finally decided to build another Wharramesque catamaran. His designs then came in two ranges: racers with about 45-degree V-hulls, and cruisers with 60-degree Vs. His only plans for a boat with the displacement we wanted had the narrow V, but a number of "Polycatistas" had told us they didn't think these hulls were faster or more weatherly by the time the cruising gear was aboard. They certainly weren't spacious. I drew a boat with the wider V-shaped section and transom sterns.

Vireo was built on a strongback so complicated that I forget exactly how it worked. Box beams were glued up first and bulkheads were attached to each side of them. Transoms and stem posts were fixed in place somehow. The hulls were planked simultaneously, and the whole monocoque structure was turned over with a crane. Throughout her construction, I used a glue called Aerolite, an English import then much discussed in the boating magazines. The end product was said to be similar to plastic resin glue, but the method of use was simpler. You bought an acid and a whitish powder. The powder was mixed with water and stored in a jar. When you needed glue, you merely painted the acid on one faying surface and the mixture on the other. You didn't have to mix each time you used, and the glue didn't stain the wood the way resorcinol had stained the inside of *Two Rabbits*.

Subassemblies that were built indoors (box beams, spars) were as well joined as if plastic-resin glued. But the boat was built outdoors, where sun and wind dried the acid at one end of a plywood sheet before the other end could be painted. This made the work more tense and rushed than if the fastest-setting epoxy had been used. It also guaranteed high maintenance for the boat after the first couple of years of her life.

Whatever glue you use, there is little to be said for the economy of building a boat outdoors. Putting up at least a roof first is always worth the trouble. Even a traditional boat, fastened only mechanically, will be twisted and checked to some extent by weather before the shielding paint can be applied.

Vireo had many of the good features of a Polynesian catamaran. She had watertight bulkheads under the forward beam in each hull. When her bows opened up on her second crossing to the Azores, these bulkheads may have made the difference between completing the trip and turning on the EPIRB. In the tops of the bulkheads, I fitted plastic deck plates which were ordinarily left open for ventilation but could be shut in an emergency. I wanted air moving through the boat at all times, and as in *Two Rabbits* I soldered up T-shaped copper vents for bows and sterns. *Vireo*'s were 3-inch diameter, and could be plugged with sponges in rough weather.

She had the meaty keels and bows of a Polycat. Wharram's keels are useful in groundings, but they are heavy assemblies even when (as in *Vireo*) they have no backbones and gluing strips behind them. The Polycat bow posts, which rake forward at 45 degrees and extend well beyond the sheer, give a romantic and even piratical look to the boats, and they make sturdy attachment points for the netting beam, anchor lines, etc. Even on *Two Rabbits* they could do some damage to a dock. But they too are heavy, and the hull shape is dictated by the shape of these posts and keels—and by the shape of the sheet of plywood.

Across the aft beam we ran a lifeline 30 inches high made of rope with two wooden stanchions. Going back and forth between the hulls in bad weather, it was always reassuring to slide one hand along this lifeline. The cabins gave a good sense of enclosure to the sides of the center deck, but the deck itself was not as wide as it had been on *Rabbits*. We made her beamier than the plans showed, but *Vireo* was built to the proportions Wharram then favored: spacing between hull centers about 38 percent of waterline length. As Jim now agrees, this is too narrow, even for a cruising catamaran. Well loaded, *Vireo* once lifted a hull in the Gulf Stream, and that should never happen. The consensus now is that hull spacing should be about half of LWL.

Center Deck

The center deck of a catamaran is the loveliest and most useful space any kind of boat can offer. No cabin that can be put on it will give so much use and pleasure as the open deck itself. In his cruising designs, Wharram specifies duckboard decks, to relieve the slamming of waves and the pressure of wind. Duckboards are also very secure to walk on, especially if made of unpainted cedar boards. Wet or dry, they are at least as non-skid as teak. Compared to nets or trampolines, duckboards are heavy; and in a boat as wet as a Polycat, they will usually contain many pounds of soakage as well. Once the pleasure of using such a deck is experienced, the weight can be borne. Even on *Rabbits*, we had a couple of folding aluminum beach chairs that we set up in port and on nice days at sea. Under an awning, we sat out many a hot noon in Bermuda or Chinco-teague or Martha's Vineyard, pitying the crews of 40-foot monohulls as they tried to make themselves comfortable in their plastic bathtub cock-pits with cabins blocking the breeze.

Below in *Vireo*, we had two bunks the width of single beds. Aft in the starboard hull was a propane stove and a dinette that seated four in a pinch. Aft in the port hull was a less useful chart room. Once beyond the smallest size, there is an inevitable redundancy of accommodation in a catamaran. Each cabin had a bucket head right aft, and we breathed easier than we often did in *Rabbits*, wondering what the crew in the galley hull was using to relieve himself.

Immediately after launching in August 1978, we shook *Vireo* down to Nantucket. The jib club, which hung from the sail and had no pivot, was soon discarded. It did make sheeting easier, but couldn't be self-tacking, because a catamaran jib must be used to get the bows through the wind. Off the wind in light air, its weight ruined sail shape. Jibing in a breeze, the club rose up like the wrath of God, ready to smite whoever was under it.

The sprit boom of the mainsail was a success, though I wouldn't want to handle one on a sail with more than a 12-foot-long foot. It had a spinna-ker pole fitting on its outboard end, as did the sprit on *Rabbits*, and for reefing or dousing sail it could be released by pulling a string. Certainly it was less bulk, clutter, and complication than a wishbone would have been. On long tacks—again like the sprit on *Rabbits*—we tacked it and set it up on the windward side of the sail. To control it, the forward end needed two lines: a snotter to hang from, and an outhaul to adjust sail camber. Some people say that sprit and wishbone booms are self-vanging, but if camber is made as full as it should be off the wind, the boom does not keep the sail from twisting. We often vanged *Vireo's* boom with a rope. The advantages of a sprit boom are that it saves the cost of a goose-neck, which might also fail at sea and be hard to mend. And a sprit boom allows foot tension to be adjusted from the center of the boat.

This was the last suit of sails Gilbert Webster made for us. Still a charming man, and still making sails for better prices than the Hong Kong lofts, he had become disorganized, it seemed, and our long-ordered working sails did not reach us until after launching day. We went to Nan-tucket without the storm sails and drifter. The drifter he cut like a genoa, and sewed the nylon panels with straight stitching. It was a flexible and useful sail, and off the wind we tacked it to the windward bow. It was not as powerful as a squaresail, but squaresails don't usually work well with Bermudan rigs, because the luffs must be too long and the wires get in the way.

Figure 7–10. Vireo under drifter at 24 degrees N 22 degrees W, the day she entered the trades.

Figure 7–11. Ruth Wharram crossing on the lifeline in bad weather.

Ruth Wharram

Early in July the next year, Ruth Wharram flew over to join us. In Cape May, we waited for a front to go through and departed for the Azores while small-craft advisories were still flying. Ruth had brought a trunkful of Polycat brochures—she never makes a distinction between work and play—and we made her stow it on the foot of her bunk. But nothing can keep this woman from being comfortable at sea. She is certainly the best ocean sailor we've ever known.

At many of the jobs that need doing, she is less than expert. She steers a good compass course, but has little feel for the wind, and doesn't make the best use of light airs to windward. She has little mechanical understanding, but she is very strong and energetic. She can jump up and down on an inflatable pump faster than the pump can recover.

On a boat in a seaway, Ruth is more relaxed and at home than on land. She can eat anything, and find it delicious. She can sleep any time, and wake any time, and still be cheerful. In a gale, when Carol and I will be moaning and popping pills and trying not to remember Hurricane Blanche, Ruth will be propped in her bunk with her hair neatly brushed, writing letters and snacking. I stagger in, clutching the hatch coamings. "There's a break in the clouds coming. I think I can get a sight, but neither of us is up to working it."

"I'll work it," says Ruth, hopping from the bunk as if from a bubble bath. "That will be fun."

Later in the gale, when the wind is down to Force 7 and the waves are no longer crashing over the cabin tops, the clouds move away. Ruth puts on her oilskins and stands for hours in the hatchway, looking out at the

Figure 7–12. Ruth's photo of a North Atlantic gale, from Vireo.

still-cascading water and the sunlight flashing off the crests. She loves the ocean, and loves everything it does.

Ruth is fearless. Carol and I do not ordinarily enjoy swimming out of sight of land, but once in light air and great heat, I decided to plunge. We were ghosting downwind at about a knot. In my bathing suit, I was standing on the transom. I looked astern, and not a boat-length away was a sizeable shark, swimming along behind us with his mouth open. Chilled more effectively than water could have done, I put my clothes back on. Not ten minutes later, Ruth was on the transom in her bathing suit. "It probably goes away now, isn't it?" she said, and dove in.

Like us, Ruth finds navigation the most rewarding hobby of ocean passages. We take sights all the time and work them with Ageton's logarithm tables, because the book is so much more compact than the H.O. 249 volumes and we got used to it on *Rabbits*. We aren't in a hurry for the answer, because after it's found, the next thing we do will probably be less interesting. We also figure our day's run from Ageton's. It will tell you the great circle distance to the destination from yesterday's noon fix and from today's, and the difference between the two is the distance made good, the day's run. The distance traveled through the water, even if accurately calculated, doesn't mean much.

Polycats like *Vireo* make good days' runs, especially on long ocean passages in regions where the wind is likely to be free. The trip we began with Ruth continued with other crew to mainland Portugal, Gibraltar, the Canaries, the West Indies, and Miami, before heading home up the Inland Waterway. On it, *Vireo* averaged 104 miles a day, which is very rare in a boat with a 23½-foot waterline. In sheltered water in the Azores, we raced her against large fleets of cruising boats, and her performance was unremarkable. But in a seaway, these boats keep going when fatter hulls are stopped by the waves. And, of course, the level platform and freedom from rolling leaves the crew rested enough to keep the boat moving. If you have just made coffee, and suddenly you must reef, you can put the coffee cup down on an unfiddled table, go on deck and do your work. You'll return to find the coffee cooler, but still sitting where it was.

A Record Run

One noontime, after a long calm on the passage from the Azores to Lisbon, the wind came in from the southwest and we began to move under main and drifter. The wind stabilized at about 15 knots, and our speed through the water brought its apparent direction to the beam. Low swells developed, with occasional but unthreatening breaking crests. We had three aboard, and after the calm we were eager to sail. For the next two days we kept her moving, heading up to pick up wind speed, bearing off to bring the speed back to course. Occasionally, we surfed on a swell, but more often we exceeded hull speed just by cutting through our bow wave with the power of the sails. At the end of 48 hours, the distance made good was 306 miles, and in both days the distance run averaged just over hull speed. It was the nicest sailing we've ever had on any boat.

Like *Rabbits*, *Vireo* had no motor, and I rowed her. But she was nearly twice the displacement, with 60 percent more wetted surface. *Rabbits* cruised at 2 knots under oar, but *Vireo* only reached 2 knots in back-breaking spurts. Cruising speed was at best a knot and a half. Up the Intracoastal Waterway from St. Lucie to Oregon Inlet, we battled with 60 drawbridges, many of them under oar. In the Waccamaw River, where the waterway goes so far inland that it's fresh and the cypress trees overhang the banks and block every wisp of wind, I rowed for hours and hours. It was a mistake. We should have bought a kicker in Miami. *Vireo* was too big a boat to do without a motor on inland waters.

The summer after the long trip, we sailed *Vireo* round-trip to Bermuda in 5.8 days going and 8.9 days returning, compared to *Rabbits'* 10 and 9.5. Bermuda is less than 700 miles from most East Coast ports, but we've always found it a nasty trip with too many weather systems to go through.

Even though *Vireo* spent 1982 on the hard, the weather continued to work on her Aerolite glue. In Madeira in 1979, we had refastened the planking to the keels with closely-spaced boat nails, and caulked her everywhere we could. Despite the potential problems caused by the breakdown in her glued joints, we departed for the Azores again in July '83 with cousin Chris Mayer for crew.

We had our first gale about four days out. Even in summer, fronts and their associated gales sweep across the North Atlantic about once a week. The farther southeast you go, the milder they become. The first gale showed us what we had been hiding from ourselves until then — how much *Vireo* had opened up. Waves that came over the cabins came *into* them also, weeping through the joints. The hulls, working in the seaway, needed bailing often, especially in the forward lockers. After the first gale, Chris and I began each day by bailing the lockers with buckets, and I suspect they filled to the waterline again in a matter of hours. If we hadn't had watertight bulkheads, we might not have kept up with it.

Below, everything was tainted with saltwater, which only *appears* to dry. Hang a blanket in the sun, and in a couple of hours it will feel dry. Put it below, and soon it will be wet again, as the salt crystals suck up moisture from the air. Some sailors have resorted to hanging a dry blanket in the rigging and beating it with sticks to get the crystals out, but that's harder to do with a mattress. In the aftermath of a later gale, when the wind had gone down but the seas hadn't yet, Chris left the companionway of the starboard hull open. We were pooped, and a wave shot the length of the cabin, floating off bedding, clothes, books, and boxes of food. It hardly mattered.

In Horta, we sold *Vireo* to friends. We told them what the problem was, and we didn't charge much for her. Remarkably, she is still afloat there, and has passed through several sets of owners. They say that for the day-sailing and weekending they do in her, she doesn't leak too badly. At least most of the water that comes down on top of her and into the bunks and stores is rainwater, not saltwater. They race her whenever they can, and when conditions exactly suit her, she wins the multihull class. She beat *Olympus Photo* (now also owned in the Azores) as recently as last year. When we arrive in Horta, it's always a treat to see her.

Hummingbird

A vireo is a green finch, a land bird that will often wander 50 miles to sea, perhaps chasing a moth or other delectable meal. On *Two Rabbits*, we often had vireos on board for the night when out of sight of land. They were a great diversion, sleeping on a stack of bowls in the galley, and we had no trouble naming our next boat *Vireo*. When we sold her, I already had some sketches of what we expected to name *Vireo II*. But the Portuguese owners told us they would keep the name as it was, in English, so we returned the compliment by naming our new boat the same thing in Portuguese, *Verdelhão*. This always makes a hit with the Portuguese fishermen we come across in New England, although it's laborious to explain to American friends. I doubt there's much merit in naming a boat in a foreign language. If you don't speak the language—we do speak Portuguese—it certainly is silly to ask people to understand something you don't understand yourself. As a name for the design, we settled on another bird that everyone knows or would like to know, the *Hummingbird*.

Our next boat was going to be a trimaran. We joked about naming her *Sirius*, after the brightest star in the sky. ("We used to have a catamaran, but then we got *Sirius*.") Catamarans are very low-resistance boats in a breeze of wind. In a puff, they have no ballast to get moving, and they don't soak up the wind's power by heeling. They accelerate almost like a car. But in light air, a catamaran must have more wetted surface and therefore more resistance than a good modern trimaran.

As an example, compare a cat and a tri, both having 25-foot waterlines, 3000 pounds displacement, and perfectly round sections. The cat's two hulls will have a wetted surface of 109 square feet, but the trimaran's one hull will have only 77 square feet. The tri's leeward float will also have

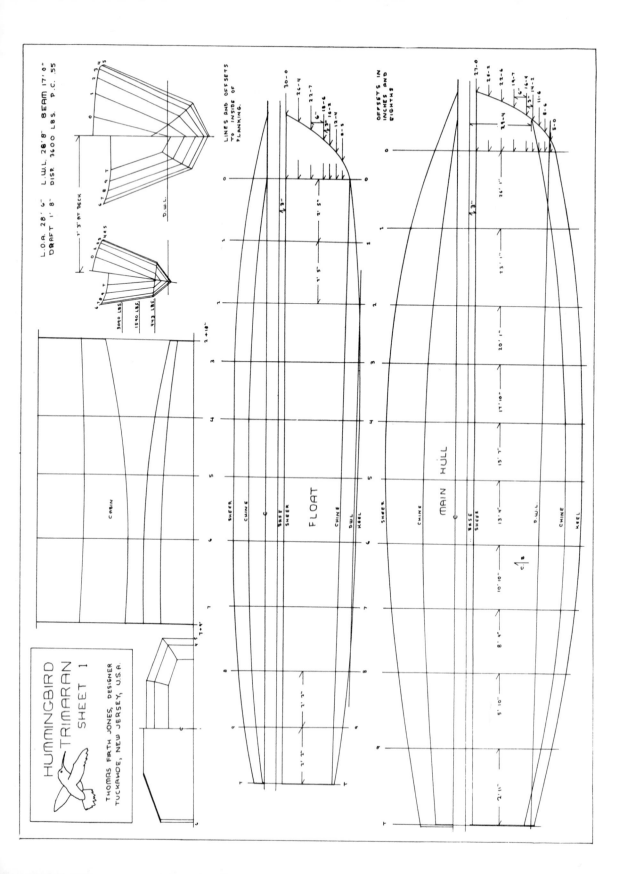

L.O.A. 28' 6"   L.W.L. 28' 8"   BEAM 17' 0"
DRAFT 1' 8"   DISP. 3600 LBS.   P.C. .55

LINES AND OFFSETS TO INSIDE OF PLANKING.

OFFSETS IN INCHES AND EIGHTHS

1' 3" AT DECK

D.W.L.

8080 LBS
1540 LBS
443 LBS

CABIN

FLOAT

SHEER
CHINE
C
BASE
SHEER
CHINE
D.W.L.
KEEL

30-0 26-4 22-7 18-6 6" 16-8 1'-3" 1'-5" 12-4 8-3

4' 3"   1' 5"   1' 5"   2' 2"   2' 2"

MAIN HULL

SHEER
CHINE
BASE
SHEER
D.W.L.
CHINE
KEEL

27-0 26-8 22-6 19-7 15" 16-4 14-2 11-6 8-5 5-0 21-4

4' 3"   26' 4"   23' 1"   20' 1"   17' 10"   15' 7"   13' 4"   10' 10"   8' 4"   5' 10"   2' 11"

C
B

HUMMINGBIRD
TRIMARAN
SHEET 1

THOMAS FIRTH JONES, DESIGNER
TUCKAHOE, NEW JERSEY, U.S.A.

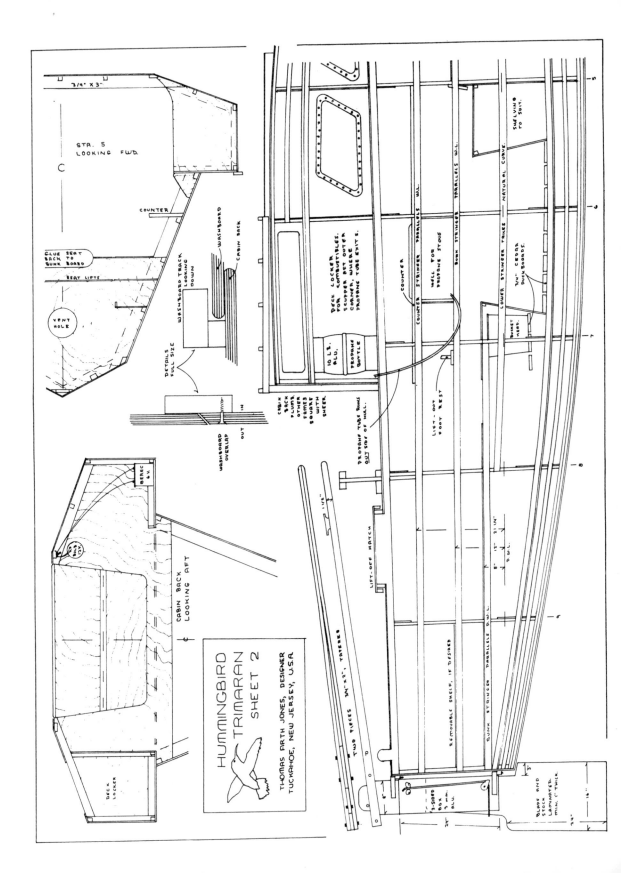

HUMMINGBIRD TRIMARAN
SHEET 2

Thomas Arth Jones, Designer
Tuckahoe, New Jersey, U.S.A.

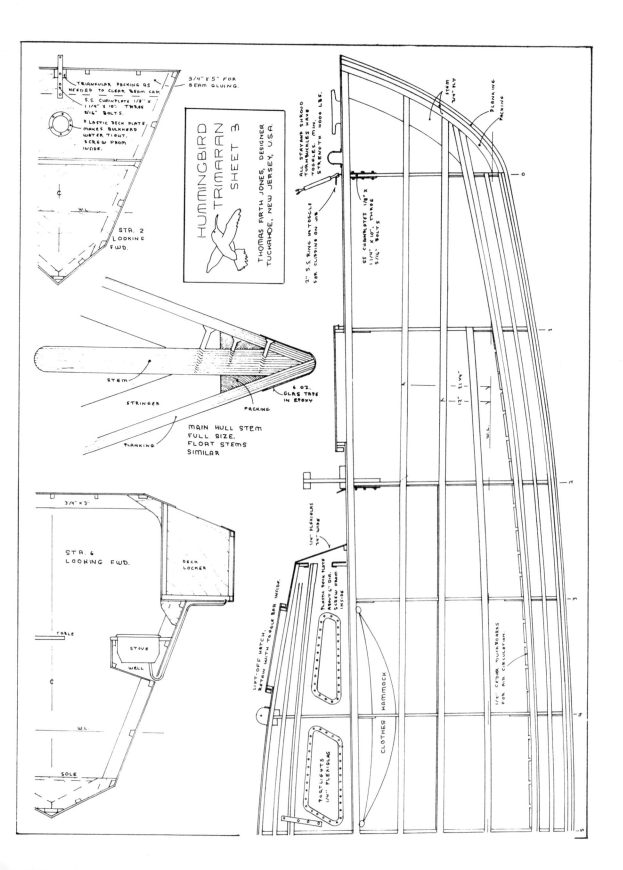

TRIANGULAR PACKING AS
NEEDED TO CLEAR BEAM CAP

3/4" X 5" FOR
BEAM GLUING.

S.S. CHAINPLATE 1/8" X
1 1/4" X 10": THREE
5/16" BOLTS.

PLASTIC DECK PLATE
MAKES BULKHEAD
WATER TIGHT.
SCREW FROM
INSIDE.

STA. 2
LOOKING
FWD.

W.L.

HUMMINGBIRD
TRIMARAN
SHEET 3

THOMAS FIRTH JONES, DESIGNER
TUCKAHOE, NEW JERSEY, USA.

STEM

STRINGER

PACKING

PLANKING

6 OZ.
GLAS TAPE
IN EPOXY

MAIN HULL STEM
FULL SIZE.
FLOAT STEMS
SIMILAR

3/4" X 3"

STA. 6
LOOKING FWD.

DECK
LOCKER

TABLE

STOVE

WELL

W.L.

SOLE

ALL STAY AND SHROUD
TURNBUCKLES HAVE
TOGGLES. MIN.
STRENGTH 4000 LBS.

STEM
3/4" PLY

PLANKING
PACKING

2" S.S. RING IN TOGGLE
FOR CLIPPING ON JIB

SS CHAINPLATES 1/8" X
1 1/4" X 10": THREE
5/16" BOLTS

0

1

12"   2 1/4"

W.L.

2

3

LIFT-OFF HATCH.
RETAIN WITH TOGGLE BAR INSIDE.

1/4" PLEXIGLAS
3/4" WIDE

PLASTIC DECK PLATE
ABOUT 6" DIA.
SCREW FROM
INSIDE

CLOTHES HAMMOCK

PORTLIGHTS
1/4" PLEXIGLAS

1/4" CEDAR DUCKBOARDS
FOR AIR CIRCULATION

4

5

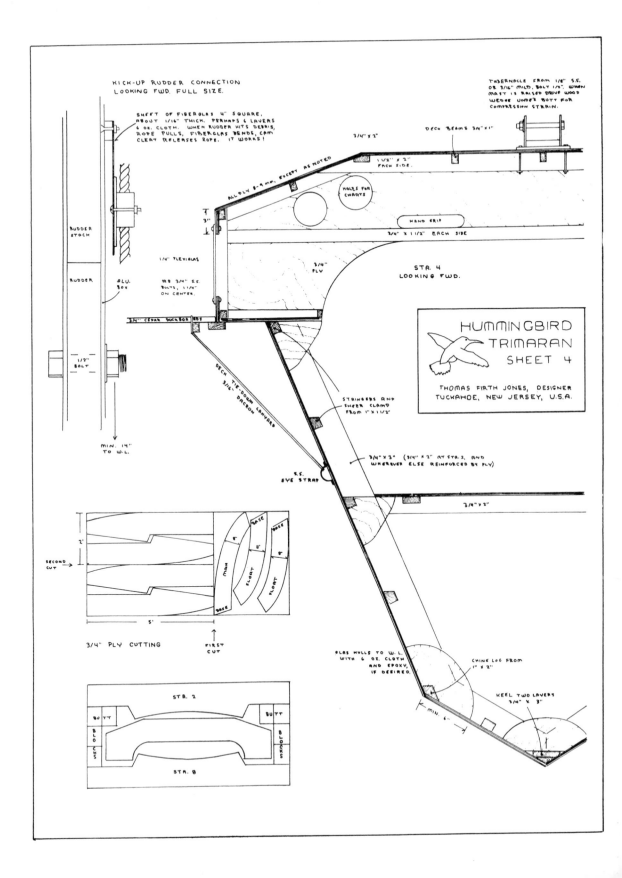

KICK-UP RUDDER CONNECTION
LOOKING FWD. FULL SIZE.

TABERNACLE FROM 1/8" S.S.
OR 3/16" MILD. BOLT 1/2". WHEN
MAST IS RAISED DRIVE WOOD
WEDGE UNDER BUTT FOR
COMPRESSION STRAIN.

SHEET OF FIBERGLAS 4" SQUARE,
ABOUT 1/16" THICK, PERHAPS 6 LAYERS
6 OZ. CLOTH. WHEN RUDDER HITS DEBRIS,
ROPE PULLS, FIBERGLAS BENDS, CAM
CLEAT RELEASES ROPE. IT WORKS!

DECK BEAMS 3/4" X 1"

3/4" X 2"

1 1/2" X 2"
EACH SIDE.

HOLES FOR
CHARTS

RUDDER
STOCK

HAND GRIP

3/4" X 1 1/2" EACH SIDE

1/4" PLEXIGLAS

3/4"
PLY

RUDDER

STA. 4
LOOKING FWD.

#8 3/4" S.S.
BOLTS, 1 1/4"
ON CENTER.

ALU.
BOX

3/4" CEDAR DUCKBOARD

HUMMINGBIRD
TRIMARAN
SHEET 4

THOMAS FIRTH JONES, DESIGNER
TUCKAHOE, NEW JERSEY, U.S.A.

1/2"
BOLT

DECK TIE-DOWN LANYARD
3/16" DACRON

STRINGERS AND
SHEER CLAMP
FROM 1" X 1 1/2"

3/4" X 3" (3/4" X 2" AT STA. 2, AND
WHEREVER ELSE REINFORCED BY PLY)

MIN. 14"
TO W.L.

S.S.
EYE STRAP

3/4" X 2"

BASE

BASE

2'

MAIN

FLOAT

FLOAT

SECOND
CUT

BASE

5'

3/4" PLY CUTTING

FIRST
CUT

GLAS HULLS TO W.L.
WITH 6 OZ. CLOTH
AND EPOXY,
IF DESIRED.

CHINE LOG FROM
1" X 2"

KEEL TWO LAYERS
3/4" X 3"

MIN. 6"

STA. 2

BUTT

BUTT

BLOCKS

BLOCKS

STA. 8

ALL PLY 8-9 mm, EXCEPT AS NOTED

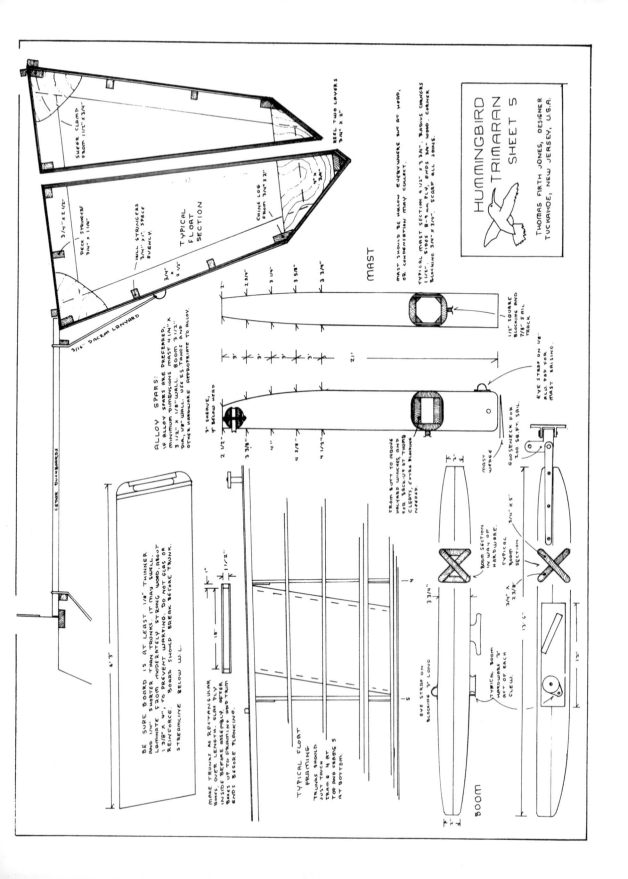

SHEER CLAMP FROM 1 1/2" X 3/4"

3/4" X 2 1/2"

DECK STRINGERS 3/4" X 1/4"

HULL STRINGERS 3/4" X 1" SPACE EVENLY

3/4" X 2 1/2"

TYPICAL FLOAT SECTION

CHINE LOG FROM 3/4" X 2"

REEL TWO LAYERS 3/4" X 2"

3/16" DACRON LANYARD

CEDAR OUTBOARDS

6'3"

BE SURE BOARD IS AT LEAST 1/8" THINNER AND 1/4" SHORTER THAN TRUNKS. IT MAY SWELL. LAMINATE FROM MODERATELY STRONG WOOD, ABOUT 1 3/8" X 4", TO PREVENT WARPING. DO NOT GLAS OR REINFORCE. BOARD SHOULD BREAK BEFORE TRUNK. STREAMLINE BELOW W.L.

1"
18"
1 1/2"

MAKE TRUNKS AS RECTANGULAR BOXES OVER LENGTH. GLAS PLY INSIDE BEFORE ASSEMBLY. OFFER BAKES UP TO FRAMING AND TRIM ENDS BEFORE PLANKING.

TYPICAL FLOAT FRAMING
TRUNKS SHOULD JUST TOUCH FRAME 4 AT TOP AND FRAME 5 AT BOTTOM.

ALLOY SPARS:
IF ALLOY SPARS ARE PREFERRED, MINIMUM DIMENSIONS MAST 4 1/4" X 3 1/2" X 1/8" WALL, BOOM 3 1/2" DIA., 1/8" WALL. USE SS TANGS AND OTHER HARDWARE APPROPRIATE TO ALLOY.

2"
2 3/4"
3 1/4"
3 5/8"
3 3/4"

MAST

3' SHEAVE, 4" BELOW HEAD

2 1/2"
3 3/4"
4"
4 3/4"
4 1/2"

3' SHEAVE, 4" BELOW HEAD
3'
3'
3'
3'
21'

FROM BUTT TO ABOVE HALYARD WINCHES AND FOR BACK-UP AT THUMB CLEATS, EXTRA BLOCKING NEEDED.

MAST SHOULD BE HOLLOW EVERYWHERE BUT AT HEAD, OR CONDENSATION MAY COLLECT.

TYPICAL MAST SECTION 4 1/2" X 3 3/4". RADIUS CORNERS 1 1/2" SIDES 8-4mm PLY. ENDS 3/4" WOOD. CORNER BLOCKING 3/4" X 3/4". SCARF ALL JOINS.

1 1/2" SQUARE BLOCKING AND 7/8" S RAIL TRACK

EYE STRAP ON 1/4" ALU PAD FOR MAST RAISING.

MAST WEDGE

GOOSENECK FOR 200 SQ. FT. SAIL.

3/4" X 5"

BOOM SECTION IN WAY OF HARDWARE.

3 3/4"

TYPICAL BOOM SECTION

3/4" X 2 3/8"

EYE STRAP ON BLOCKING 9" LONG

13' 6"

12"

TYPICAL BOOM HARDWARE 9" AT OF EACH CLEW.

EYE STRAP ON BLOCKING 9" LONG

BOOM

HUMMINGBIRD TRIMARAN
SHEET 5

THOMAS FIRTH JONES, DESIGNER
TUCKAHOE, NEW JERSEY, U.S.A.

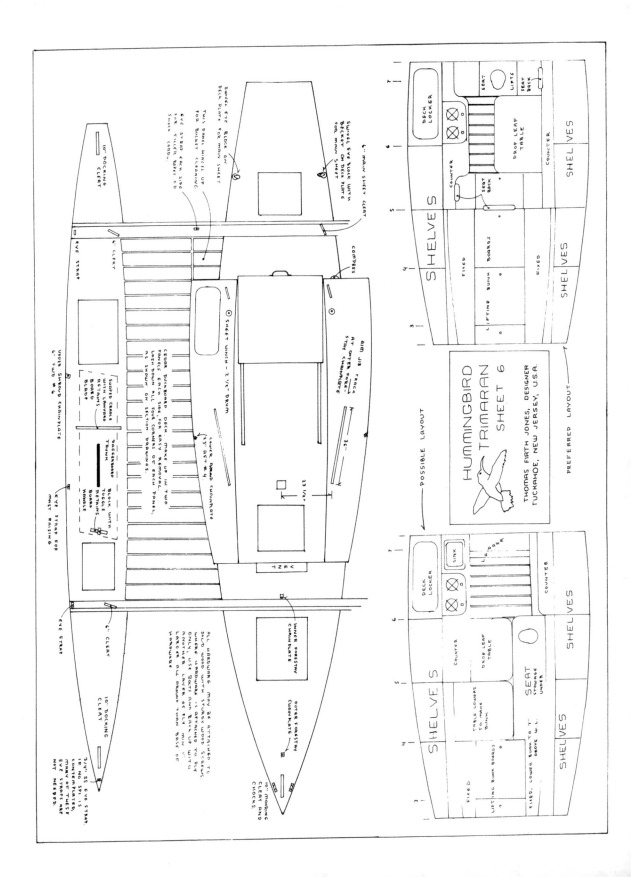

HUMMINGBIRD TRIMARAN

SHEET 6

Thomas Firth Jones, Designer
Tuckahoe, New Jersey, U.S.A.

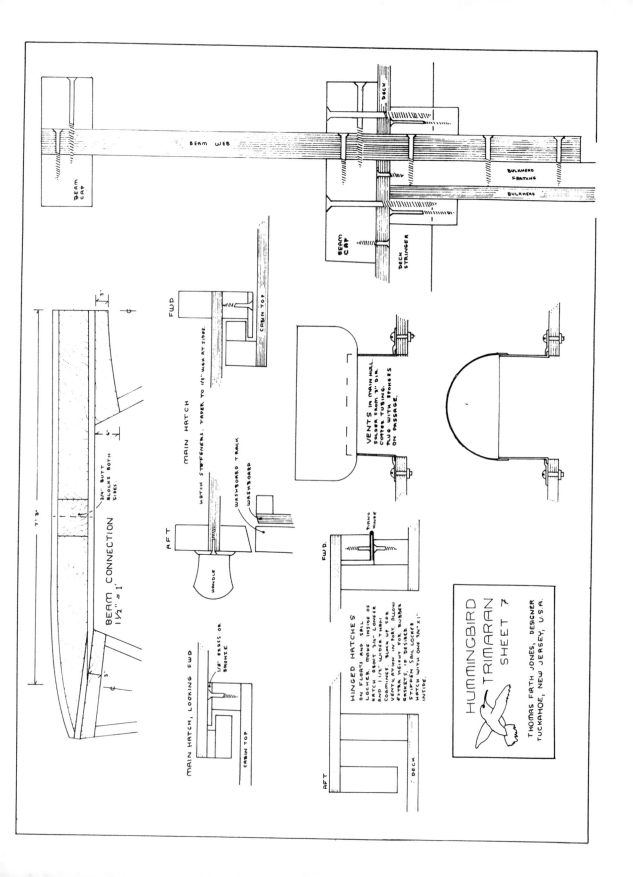

BEAM WEB

BEAM CAP

DECK

BEAM CAP

DECK STRINGER

BULKHEAD FRAMING

BULKHEAD

BEAM CONNECTION
1½" = 1'

3/4" BUTT BLOCKS BOTH SIDES.

7'-3"

3"

6"

3"

FWD

CABIN TOP

MAIN HATCH

HATCH STIFFENERS. TAPER TO 1/2" HIGH AT SIDES.

AFT

WASHBOARD TRACK

WASHBOARD

HANDLE

MAIN HATCH, LOOKING FWD

1/4" BRASS OR BRONZE

CABIN TOP

VENTS IN MAIN HULL. SOLDER FROM 3" DIA. COPPER TUBING. PLUG WITH STONERS ON PASSAGE.

FWD.

PIANO HINGE

AFT

DECK

HINGED HATCHES

ON FLOATS AND SAIL LOCKER. MAKE INSIDE OF HATCH ABOUT 3/4" LONGER AND 1 1/2" WIDER THAN COAMINGS. BLOCK UP FOR VENTILATION IN PORT. ALLOW EXTRA HEIGHT FOR RUBBER GASKETS, IF DESIRED. STIFFEN SAIL LOCKER HATCH WITH ONE 3/4" X 1" INSIDE.

HUMMINGBIRD TRIMARAN
SHEET 7

THOMAS FIRTH JONES, DESIGNER
TUCKAHOE, NEW JERSEY, U.S.A.

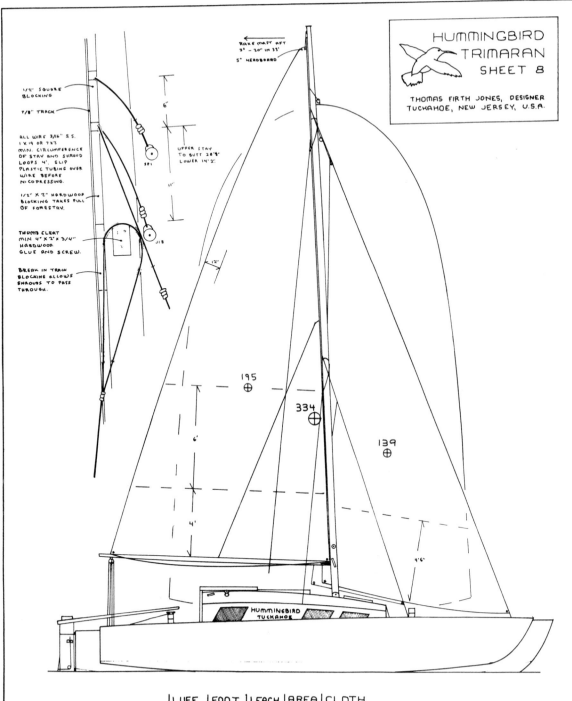

HUMMINGBIRD
TRIMARAN
SHEET 8

THOMAS FIRTH JONES, DESIGNER
TUCKAHOE, NEW JERSEY, U.S.A.

RAKE MAST AFT
3" - 20" IN 33'
5" HEADBOARD

1/2" SQUARE
BLOCKING

7/8" TRACK

ALL WIRE 3/16" S.S.
1 X 19 OR 7X7
MIN. CIRCUMFERENCE
OF STAY AND SHROUD
LOOPS 4". SLIP
PLASTIC TUBING OVER
WIRE BEFORE
NICOPRESSING.

1/2" X 2" HARDWOOD
BLOCKING TAKES PULL
OF FORESTAY.

THUMB CLEAT
MIN. 4" X 2" X 3/4"
HARDWOOD
GLUE AND SCREW.

BREAK IN TRACK
BLOCKING ALLOWS
SHROUDS TO PASS
THROUGH.

6"

UPPER STAY
TO BUTT 28'4"
LOWER 14'2"

11"

SPI

JIB

12'

195

334

139

6'

4'

4'6"

HUMMINGBIRD
TUCKAHOE

	LUFF	FOOT	LEACH	AREA	CLOTH
MAIN	30'	13'	31'6"	195'	7 OZ. DACRON
JIB	28'6"	11'6"	24'6"	139'	" " "
TRISAIL	14'	6'6"	15'	46'	" " "
STORM STAYSAIL	12'6"	5'9"	10'7"	31'	" " "
SPINNAKER	30'	18'	———	500'	3/4 OZ. NYLON

Figure 7–13. Verdelhão *in the Tuckahoe soon after launching day.*

Figure 7–14. Verdelhão *under pole-less cruising spinnaker.*

Figure 7–15. Original lashed-on floats allowed too much play, and were replaced by glued and screwed plywood I-beams.

some wetted surface, but in light air it will be nothing approaching 32 square feet, and the windward float will be out of the water. The problem with the trimaran is that the main hull must still be fine to exceed hull speed in stronger winds, and the structure of a tri is likely to weigh at least as much as a cat, so in any given size, a good tri will carry less payload than a good cat.

Carol and I wanted to keep our new boat under 30 feet. Length is not the only measure of a boat's size, but it does have some relation to building time and cost, and certainly to licensing and docking charges which are inevitably based on length alone. The math for *Vireo* presupposed an 1800-pound payload to get three people across an ocean. Such a payload is hardly possible in a fine-hulled trimaran under 30 feet.

A tri became possible for us when we decided to stop taking crew on ocean passages. In *Vireo*, we had usually carried a third person, and this made for a very easy and pleasant schedule: 3 hours on and 6 off, so everyone got most of a night's sleep, and the watches rotated each day. Even in *Rabbits*, we had often taken crew, though not on the way home from Bermuda, thank God. Some of our crew had been better than others, though none had been as good as Ruth. The majority of them, no matter what their inshore sailing experience, had suffered from "sea shock" in deep water.

James Wharram says that the sufferer from sea shock often "is, or wishes to be, in absolute control of his 'tools' and life pattern. On land, he molds life around him. He expects the same pattern at sea." When the sea begins molding *his* life, his reaction will depend on his temperament. He can become seasick, quarrelsome, somnambulistic, or all of these things in turn. Carol and I have had our own troubles with the problem but, by 1983, we had had enough of other people's sea shock. We had made a number of passages double-handed, and knew it could be done. We decided this was preferable to coping with crew.

With one less person and his gear, water, and food aboard, and with other small weight economies such as an aluminum instead of a steel propane bottle, we thought we might get by with a 1200-pound payload. I had already spent many wonderful days talking and sketching trimarans with Vance Buhler, who builds head-boat sailing catamarans (sometimes called "cattlemarans") in St. Vincent, West Indies. Few people understand the compromises of yacht design better than Vance, and no one articulates them better face to face. Lamentably, he isn't much of a writer, being more interested in the word processor than the words.

Twenty years ago, Robert B. Harris wrote a good book about trimaran design, but it's now out of print. Study plans also contain much information if read carefully, and they cost very little. In drawing the *Hum-*

mingbird, I kept beside me the study plan of a boat the same size but much racier—Locke Crowther's *Buccaneer 28*. I also had complete plans of a smaller boat, the Cross 24, higher resistance but with a proportionately bigger rig to compensate. Perhaps not much of *Hummingbird's* design is original, but not much is wrong either. She is exactly the compromise we wanted between speed, comfort, and cost. Because they don't heel, multihulls are much easier to design than monohulls. Any experienced sailor who can do the numbers and isn't besotted with his own originality can make an acceptable job of it.

Arthur Piver

Designing the first trimaran was trickier. Good catamarans began to appear soon after World War II, designed and built by young men home from the Pacific who had seen and admired the local boats. A number of people also built and tested trimarans, and wrote about their boats for limited audiences such as the Amateur Yacht Research Society, but no one got the proportions exactly right—length to beam of the hulls, hull spacing, float volume. Not until 1958 did one AYRS member, Arthur Piver, editor of a San Francisco insurance newspaper and with as little training in yacht design as James Wharram, put together a boat so good that it has been the standard of trimaran design ever since.

Nugget was such a stroke of genius, such an extraordinary combination of dogged work and sudden insight, that in later years Piver himself was at a loss to explain it. "Small boat design is more of an art than a science," he wrote in his usual self-conscious, self-congratulatory manner, "and we have here a new expression of the designer's art." Piver was at least as much a publicist as a designer, and in the next 10 years he sailed his trimarans across oceans any number of times, and wrote three fascinating but infuriating books. He sold so many sets of plans that, according to his one-time protege, Jim Brown, he may have made more money at yacht design than anyone else ever did.

"I thought he was a very interesting and complex person," says Ed Cotter, who wrote a couple of books about early multihulls. "Others thought he was a nut, because he oversold some of his ideas. His manner was to overwhelm a person, and possibly make some outlandish claims, which were later used to discredit him." He sure pulled Jim Wharram's chain. Today, 23 years after Piver's death, Wharram will still use his column in the Polycat journal to fulminate about Piver.

Piver was a very good sailor. He had grown up on boats, and he was quick and strong. He didn't get seasick, and he wasn't afraid. Almost from the start, he planned to establish his boats' reputation once and for all by winning the Observer Singlehanded Trans-Atlantic Race; but for a number of reasons—mostly his own gregariousness—he never crossed the starting line. By 1968, other designers were producing more sophisticated and faster trimarans, though they didn't stray far from Piver's essential proportions. The organizers of the race that year told him that, although he had been entered in two earlier races, he would have to qualify for this one with a 500-mile solo passage. Piver's racing boat was already in England. He cast about for a boat close at hand, and chose a 25-foot motorsailer of his design, and not a well-built example.

He sailed from Sausalito on March 17, bound for San Diego. The weather was forecast to be good, and it remained good for two weeks. The temperature was close to 60 degrees, the sky was clear, and the boat was close hauled to a moderate westerly wind as she sailed out under the Golden Gate Bridge. That was the last anyone ever saw of Arthur Piver. Robert Harris discusses what may have happened: The boat may have broken up, or been hit by a steamer; or Piver may have fallen overboard, because he always relied on his own wits and agility, not on safety gear. Despite a five-day Coast Guard search, covering 50,000 square miles of ocean, no wreckage was ever found, and we will never know.

The floats of Piver's trimarans were deeply immersed, with 60-degree V-sections something like a Polycat hull, canted slightly outboard, and attached to each side of the main hull. Piver defended his floats, saying that only they would give an acceptable sea motion. Maybe so, but his trimarans had no less wetted surface than catamarans the same length. The main reason other designs began running away from his was that their floats barely touched the water at rest, and in the lightest wind one float came out of the water altogether.

Twenty years ago, it was the wetted surface advantage of trimarans that made them the race winners. Then catamarans began using such big rigs that they sailed most of the way across oceans with one hull out of the water, like Hobie Cats with no float to drag along. Cats began winning. Then trimarans began using even bigger rigs, and boats began sailing with not just a float out, but the main hull out of the water, too. Main hulls shrank, floats expanded, and rudders moved out to the floats. Tris, with their longer righting arm, began winning again. All this is entertaining, of course, but has little application to cruising boat design.

Designing a Trimaran

Length-to-beam ratio of any multihull must not be less than 8 to 1, and draft not more than half of beam, or the hulls will not escape from their own wave train with the low power that sails generate. Thus a perfectly round-bottomed trimaran of these proportions, 26 feet 8 inches on waterline with reasonably fine ends, can't displace more than 4000 pounds.

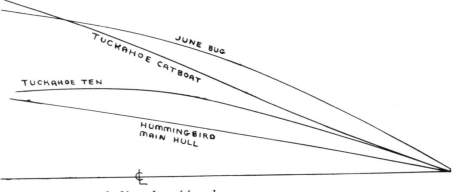

Figure 7–16. *Bow half-angles of four boats.*

The main hull sections should be identical, or nearly so, for several stations amidships, and the entry and run somewhat abrupt. This gives a P.C. of .55 or better, dampens pitching, and gets the load through the water faster than would finer ends and a larger midship section. This allows the cabin sole to be lower farther aft, giving more headroom. Transom shape should be a compromise between too fine, which may allow pitching, and too broad, which in a following sea in extreme circumstances might cause a light multihull to pitchpole. Bow half-angle is likely to be very fine in an 8-to-1 hull; *Hummingbird*'s is 11 degrees.

Float design follows the same principles, but even more extremely, because these hulls are not deeply immersed unless the boat is going very fast. Prismatic should be at least .60, and length-to-beam will be more like 20 to 1. Floats should be nearly as long as the main hull, and each should have enough buoyancy to support at least 150 percent of the weight of the whole boat. Their transoms should be very fine, and their overhangs considerably longer than the main hull's, to keep down drag in varying states of loading. To shorten the connecting structure, they should be canted outboard to the boat's expected angle of heel to windward.

Hull spacing in cruisers should be exactly what Piver used in *Nugget*, giving an overall beam about 60 percent of overall length, or no wider than a good modern catamaran. Wider spacing gives more power to carry sail, but also a jerkier ride and more strain on the beams. Even a narrow tri will have far more righting arm than a cat, because the floats cant out. The limiting factor on how narrow a tri can be is not righting arm, but rather the wave interference between the hulls.

For building material I never considered anything but plywood. *Hummingbird* has about 1000 square feet of skin, and the finish that the factory puts on the plywood panel would be reason enough to choose it over any other material. I used lauan marine. Khaya would have raised the materials cost of the boat about 20 percent, and would have been an even better choice, though the lauan has given no trouble in seven seasons.

Plywood requires chines, unless the hull is very long and narrow, and can be tortured from thin sheets. Chines slow the boat. *Hummingbird* has 3 percent more wetted surface in her main hull than a perfectly round-bottomed hull of the same length and displacement. However, chines do not cause the wave-making resistance in a narrow multihull that they do in a fat monohull. In any case, building our trimaran took seven months of my full-time work, and that's near my maximum concentration span for any one project.

The beam connections were at first Dacron lashings, but we found that no matter how tightly we cinched the lines, they would always stretch a little. The stretch made play, which augmented the shock loads and increased the stretch. She held together for her first 1000-mile cruise, but the following winter I cut the decks and glued and screwed plywood I-beams to the bulkheads, making the boat monocoque. She should have been that way in the first place.

Hummingbird has a daggerboard which is shifted from one hull to the other when tacking. This is a nuisance, and in narrow water we don't do it, and settle for inferior performance on one tack. A second daggerboard would spare a little of the nuisance, but would weigh the same 23 pounds as the first and would have to be subtracted from payload. We considered twin centerboards, but these and their trunks would have been heavier, and the trunk slots would have weakened the hulls and made turbulence. I always think about putting a centerboard in a boat, and always wind up with a daggerboard. We never considered a single centerboard in the main hull because that would have dictated the accommodation in a boat this size.

Vireo went to weather better than *Rabbits*, mostly because she was bigger. Waves are the same size and wind the same strength, no matter what size the boat. In a part of the ocean with negligible current during the last four days of *Vireo's* second passage to the Azores, she was dead-headed but always moving, sometimes single reefed and sometimes with plainsail. She averaged 60 miles a day made good, or 2½ knots, compared with *Rabbits'* 1½ knots.

We were very interested to try *Hummingbird* out to windward. Without the daggerboard, she definitely made progress, tacking in 90 or 100 degrees or whatever you like. She made some visible leeway, but threw a nice wake, and gave a lively, interesting feeling. In truth, she didn't do too much worse than a Polycat in the same conditions. When we put down the daggerboard, the wake disappeared. It was caused by dragging the hulls sideways through the water. Speed appeared to be less, but believe me, it wasn't. In a decent wind and seaway, *Hummingbird* will make good better than 4 knots to windward. This may not seem spectacular, but she will pick off all the 40-foot monohulls on the upwind leg of a race. In ghosting conditions, she will keep making progress toward her destination when boats with inadequate leeway prevention are going straight sideways.

Multihull Rigs

Like a monohull, a multihull can have any rig you like: ketch, barkentine, etc. (In Scituate, Masschusetts, we once saw a ship-rigged, 10-foot rowboat. The squaresails were pillowcases. One youngster sailed it down the harbor, and another in a small outboard towed it up again. They seemed to be having a good time.) But in multihulls, the evolution is toward a sloop rig, because the boats go faster than monohulls and bring the apparent wind farther forward more of the time. The same kind of evolution, to more extreme degree, has happened in iceboat rigs.

Some of the older, heavier, narrower multihulls that are still in production have masthead sloop rigs. On the Prout *Snowgoose*, for example, the jib is so big that the main is little more than a trailing flap for it. Prout makes this work pretty well, because along with the weight and narrowness comes the structural stiffness needed to keep the forestay taut. Modern designs with lighter structures and wider hull spacing do not have this stiffness and do better with fractional rigs. The mainsail gives most of the driving power, and the jib does little more than direct the air across it. Inshore, a Bermudan main can profitably be fully battened, which can increase the area as much as 50 percent. Offshore, full battens are probably too troublesome, though sailmakers are working to improve them.

For the first several years we sailed *Vireo*, the balance of her sailplan was similar to a monohull sloop. Close hauled in any wind above 12 knots, she developed strong weather helm. Sometimes we reefed just to spare the helmsman's arm. Finally, Locke Crowther explained to me that the balance problem is different in multihulls. The mono heels and produces weather helm from her asymmetrical shape under water, and from the wind pressure on the rig being way out to one side. The solution is to shift the center of effort by reefing the main and continuing with the genoa. The multi produces weather helm from the sail area forward pushing the fine bows down into the water and moving the center of resistance forward. The solution is to move the whole rig farther aft and, if need be, reef the headsail first.

I moved *Vireo*'s mast aft a foot, and moved the forestay chainplates half-way down the bowposts. The weather helm diminished, and the boat became far easier to manage. In *Hummingbird*, I set the mast far aft to begin with, and set the forestay well in from the bow. There is almost no lead of effort over resistance. Balance is good in a variety of wind strengths.

Initially, she had a 30-foot mast and a mainsail of less than 2-to-1 aspect ratio. When these sails were worn out after 13,000 miles, I made a new mast 3 feet longer with 2½-to-1 mainsail and a taller and narrower jib. Sail balance was maintained by raking the mast farther aft. The new rig is about the same area as the old, and no more powerful off the wind except that it allows another panel to be added to the spinnaker. On the wind, it makes a very much faster boat, and the sails are easier to handle, too. Although the taller mast is harder to raise, it is still considered conservative by modern standards.

The plans show two accommodation layouts for *Hummingbird*, and more could be thought of. *Verdelhão* has the diners facing fore and aft, and a bunk that needs no breaking down. We do without standing headroom below. Many a 28-footer has it, but the cost in weight, windage, and center of gravity comes pretty high. At sea, the prudent sailor spends most of his hours sitting or lying down. In port, we have standing headroom in the hatchway, which is almost 3 feet square. Belowdecks headroom is 5 feet 5 inches, which suits Carol fine.

The cabin space is divided almost evenly between the saloon and the wide, double bunk. Headroom over the bunk is nearly 4 feet, except under the mast-support beam. The bunk is low and near amidships where the motion is least.

A curtain divides the bunk from the saloon. Heavy stores, such as water and food, go under the bunk, and clothes go into hammocks above and outboard of it where they catch some air and sun. Shelves in the overhangs contain books, navigation equipment, and health and beauty aids. Holes in the framing gussets take rolled charts. A collision bulkhead separates the bunk from the sail locker forward.

The single bunk aft is accessible through a deck hatch or through the saloon. Despite the portlight in the hull, it's a miserable place, with hardly more than 2 feet of headroom. Ruth Wharram did once make herself at home in it for two weeks, but she's tougher and cheerier than most.

The saloon is 6 feet long. Two people can sit at each end, with a light poplar table between them. Even in cloudy weather the saloon seems airy and bright. The paint job — varnish below the sheer and white above it — contributes to this atmosphere. There is storage space under each seat, and a bucket head under the aft one.

Figure 7–17. Vent in cabin front, with plexiglas baffle. Also used on Vireo, Scampi, *and* Elegant Slider.

Figure 7–18. Well in overhang for propane bottle, outboard, and other combustibles.

The cabin overhangs the hull 2 inches forward, 9 inches amidships, and 13 inches aft. At the cost of some complication, weight, and expense, this creates visual space below and gets the galley shelves out of the cook's face. Our caged pet mice find a home there, though they don't much like sailing. Lately, we have had another freeloader: a tiny blue-winged parrotlet. Parrots enjoy the motion and ruckus of sailing, and have no illusions about privacy or respect. The overhangs also knock down spray, and the *Hummingbird* has proven a very dry boat. But if extended too far outboard or too far forward, overhangs will pound and, in extreme cases, detach themselves from the hulls.

We have no cockpit, and haven't on any of our cruising boats. Cockpits are for daysailers. On passage, we don't steer when we can help it. We're often able to make our boats steer themselves with sheet-to-tiller arrangements, though the Polycats do this better than *Hummingbird*. When you *must* steer, it's nice to have part of your body inside the hull to be in some communication or at least visual contact with the off-watch, perhaps to kibitz a game of solitaire. We sit on the deck, with our feet on a stick that rests on the stringers. Tacking, Carol steers and I stand on the seat in front of her, rattling the jib winches ostentatiously. In port, we sit out on the side decks, though they aren't as nice as the center deck of a catamaran.

Our electrical system consists of a drycell battery with a toggle switch for the compass light. It has been troublesome enough to make us glad it isn't more complicated. We have no plumbing system, but carry our water in gallon bottles and dump it overboard after use, instead of pouring it through a hole in the hull. We have no other systems either. We go to sea to sail, and leave the systems at home.

We have tried a number of powering arrangements on *Verdelhão*, and never found a good one. A catamaran can have an outboard on the aft beam, some feet away from the pitching of the sterns and protected by the hulls from wakes and spray. On a trimaran, an outboard can only be on the main-hull transom, or at the junction of the main hull and the aft beam (this boat is far too small for an inboard). In either place, it is subject to frequent wetting. In a flat calm, a 3-horsepower outboard at half throttle will drive a *Hummingbird* at 4 knots. We sometimes use one in port. We have not found any way to motor her in a seaway, but the *Hummingbird* sails very well.

Performance

Comparing numbers with some other boats about her length, sail area-to-wetted surface ratio (S^A/wS) should be at least 2.25, says *Skene's*, though multihulls can often get away with less. As said before, Bruce number should be at least 1.00. All numbers are with the boats loaded to their marks and the board, if any, down:

	SA/WS	BRUCE NUMBER
Piver Nimble 30	1.63	1.00
Vireo 27	1.98	1.10
Cross 28	2.15	1.19
Hummingbird 28½	2.62	1.18
Crowther Buccaneer 28	3.83	1.35

These boats actually perform very much as their numbers suggest, though *Vireo* and *Nimble*, with their inefficient leeway prevention, do badly to windward. Off the wind in fresh conditions, the *Hummingbird* is no faster than *Vireo*. We have no speed-measuring equipment (which is one of the reasons we go fast), but we think we saw 15 knots in *Vireo* once or twice, and never more than 12 in *Verdelhão*. Perhaps we're older and more cautious now, and certainly the bigger, drier boat gives less sensation of speed. But our Polycats both exceeded hull speed without our noticing, and the wider main hull of *Hummingbird* sends little signals as she crosses the barrier—trim changes a couple of degrees, and the wake becomes clamorous. We've had some wonderful short runs in her: 92 miles up Chesapeake Bay in 11 hours; 120 miles from Montauk to Barnegat sea buoy, pinched and reefed, in 18 hours. With three people aboard on a short ocean passage, we could push her harder, and we might see a 200-mile day. Best so far was last August when we came home from Montauk to Ocean City, 162 miles in just over 24 hours.

It is not in her top speed, but in her average speed that the *Hummingbird* excels. No doubt a Buccaneer 28 with its low-resistance hulls and enormous, efficient rig, could give her a good dusting in any condition, but the Buccaneer is an inshore boat. We just as effectively dust the average 40-foot monohull when we come up on one. They look like they're dragging nets. We sail three miles to their two, and often pass them under sail when they're motorsailing. Crossing Nantucket Sound with a 10-knot breeze just forward of the beam, we had no trouble getting past the 90-foot *White Hawk* with her vaunted Herreshoff look and her "modern" underbody. Why do people put up with these clunkers?

We took our *Hummingbird* on a 10,500-mile cruise in 1987–88, visiting the Azores, Senegal, Cape Verdes, Cayenne, and the West Indies. We had a number of accidents, such as taking the bottom out of the main hull on a lava reef, and we spent almost as much time working on the boat as sailing her. When she was whole, she averaged a reliable 125 miles a day. This is a 20-percent improvement on *Vireo*, which is good in a boat only 1½ feet longer. But on these long trips, you don't do too much upwind sailing, and there the improvement is 60 percent.

Our most frequent accident on the trip was sheering off the rudder or breaking the rudder hardware. We just hadn't realized how much more junk was floating in the ocean than had been eight years before. Containers, barrels, skids, and logs now litter the North Atlantic, and because we were usually going fast, we hit them hard. We finished the trip with a gudgeon welded up from mild steel by a motel handyman in Dakar. First modification, when we got her home, was to make a kick-up rudder for *Verdelhão*. Unfortunately, the amount of junk too small to damage a boat has also grown exponentially in recent years: light bulbs, bottles (often deliberately corked), and plastic trash come up on every other wave. In calms, black balls of crude oil spot every square yard of the ocean surface.

We mean to keep our *Hummingbird*. Capitalism makes all of us itchy for a new boat, car, house, job, mate; we must constantly resist this itch or our life becomes a misery of unfulfilled expectations. Take another look at your own boat. Maybe she ain't so bad. But if I were building again for the coastal cruising we now have in mind, it would be a cat, not a tri. It would be something like the catamaran Bill Weisbrod was drawing, but without the complications and puzzles. It would be smaller than *Verdelhão*, lighter and less teetery, easier to haul and launch from our muddy beach. It would have a mast that I could raise without tackles. For a building method, I was thinking. . . . No, I wasn't thinking, and won't.

We like cats better than tris. We love the speed of the tri, especially in light and moderate air, and the quick tacking. We love showing *White Hawk* and the Fat-Assed Forties where to head in. But many of these pleasures would still be available, though in diminished measure, in a catamaran with carefully designed hulls and a good board. We prefer the motion of a catamaran, which is quicker but shorter, both at sea and in harbor. We find that the 10 degrees a trimaran can heel going to windward is too much for us. And we really don't like the way a modern tri will walk from one float to the other in wakes at anchor; or how, when you're underway, the windward float will periodically touch the sea and throw a jet of water at the main hull.

Cats are more habitable. Not only is their center deck the nicest space on any boat smaller than an ocean liner, but the two separate living spaces give you somewhere to go, something to do, something to think about. In layout and perhaps even in finish, they should be as different from each other as possible. There is no reason, beyond the old bugbear of symmetry, why the two cabins should mirror image each other.

There you have it, shipmates, as William Atkin used to say, another vessel to ponder.

Afterword: Designing and Building Your Own Boat

Curiously, when a neophyte decides to design a house, he chooses an arbitrary shape first and tries to shoehorn the rooms into it. Conversely, when he designs a boat, he draws the accommodation first, and optimistically sketches the hull shape around it later. This is exactly the opposite of what should be done. A house doesn't move, and can be virtually any shape. A boat, however, must cut through the water, as well as contain amenities. The place to start with boat design is with numbers, and specifically with weights.

Nearly all formulas and parameters needed for boat design are contained in *Skene's*, but they are so compacted that they are not always understood, even at third or fourth reading. Chapelle's *Yacht Designing and Planning* goes slower, but covers less ground. Weston Farmer was bombastic and kitschy, but often valuable. Some magazine articles contain golden snippets, but they are too soon thrown out or forgotten. It helps to keep a file of designs and ideas, and to read as widely as possible. Ted Brewer's *Understanding Yacht Design* was the text that got me started, and it's still a good one. The problem of getting the payload through the water easily is not easily solved.

Few kinds of boats demand an entirely original design, and few designers are capable of an entirely original solution. Many of the boats in this book owe a great deal to other designers, and I suspect that other designers are no less beholden than I am, though some are more guarded in naming their sources. Don't hesitate to borrow. But don't imagine that you can take the lines of a Rushton kayak, double them, add a pilothouse with standing headroom and a V-8 engine, and arrive at a useful boat. One designer-builder I knew of spent his life as a tugboat crewman. For retirement, he conceived and built an unballasted, flat-bottom sloop with a telephone pole for a mast. Not long after launching day, she fell over in her slip, and he spent his golden years elsewhere. Do your math.

As for building, there are many good texts. Steward's recently revised *Boatbuilding Manual* may be the best of them, but Chapelle and Verney and the Gougeon Brothers also are useful. Visiting boatshops also helps, and will often be tolerated if you don't stay too long and confine yourself to asking, not telling. In general, the amateur builds better than the professional, but takes incredibly longer at it. Start early, but start with a feasible project, and keep moving. Once you have started, do not add complications.

Professional designers and builders may tell you that you can't do, on your own, what they do for a living. This is not true, of course, though their motives for saying so are not hard to understand. William Garden will allow that building a boat is "as close as one man can come to the nearly complete creation of something with a personality and life." Designing as well as building the boat comes a step closer yet. Start reading up on it.

Selecting a boatbuilding material is not as important as some people think. Often a design can be carried far along before materials and methods are settled on. Apart from idiosyncratic materials like tortured plywood, and heavy ones like metal and ferrocement, a design can often progress almost to completion before a hull material is chosen. It is usually a mistake to start to design a fiberglass hull, or a carvel-planked one. Often designer-builders who do so can work with only one material, such as plywood, or trust one material, such as steel.

My own preference is for plywood. Despite its shape limitations, plywood offers light weight, stiffness, strength, low price, and a factory-made finish available in few other materials. It also biodegrades at a satisfactory rate, leaving less clean-up for the next generation. But there is no one material or method best suited to boatbuilding. Try to become easy with all the possibilities, and postpone the decision until it is suggested by the design itself.

Index